SUSTAINABLE DEVELOPMENT
AS A CIVILIZATIONAL REVOLUTION

Sustainable Development as a Civilizational Revolution

A Multidisciplinary Approach to the Challenges of the 21st Century

Artur Pawłowski

Faculty of Environmental Engineering, Lublin University of Technology, Lublin, Poland

CRC Press
Taylor & Francis Group
Boca Raton   London   New York   Leiden

CRC Press is an imprint of the
Taylor & Francis Group, an **Informa** business

A BALKEMA BOOK

CRC Press/Balkema is an imprint of the Taylor & Francis Group, an informa business

© 2011 Taylor & Francis Group, London, UK

Typeset by Vikatan Publishing Solutions (P) Ltd., Chennai, India
Printed and bound in Great Britain by Antony Rowe (a CPI Group Company), Chippenham, Wiltshire

Published by: CRC Press/Balkema
                P.O. Box 447, 2300 AK Leiden, The Netherlands
                e-mail: Pub.NL@taylorandfrancis.com
                www.crcpress.com – www.taylorandfrancis.co.uk –
                www.balkema.nl

ISBN: 978-0-415-57860-8 (Hbk)
ISBN: 978-0-203-09338-2 (eBook)

Table of contents

About the Author   vii

Introduction   ix

CHAPTER 1: The evolution of the idea of sustainable
development in history   1

1. Worldwide perspective   1
    1.1 Early initiatives   1
    1.2 Contemporary later advances   7
2. European perspective   29

CHAPTER 2: Theoretical basis for sustainable development   37

1. The notion of sustainable development   37
2. Hierarchy of planes   45
3. The principles of sustainable development   47
4. Indicators of sustainable development   53

CHAPTER 3: Philosophy, religion and environmental education   57

1. Eco-philosophy and the ethical plane of sustainable development   57
2. Religion and sustainable development   72
3. Ethics in practice: Ecological attitude and education   76
4. A new research field: A philosophical audit   81

CHAPTER 4: Level II of sustainable development: Ecological,
social, and economic considerations   85

1. Ecological plane   85
    1.1 Natural environment   85
    1.2 Changed landscape   92
2. Social plane   101
    2.1 Social environment   101
    2.2 Cultural landscape   104
    2.3 Urbanization and Healthy Cities   111
    2.4 North vs. South   116
3. Economic plane   123
    3.1 Traditional economy vs. ecological economics   123
    3.2 Economic instruments for protection
         of the environment   131
    3.3 Responsible business and environmental
         systems of management   133

3.4   Financial security for introducing sustainable
        development   138

CHAPTER 5: Level III of sustainable development: Technical,
legal and political considerations   145

1. Technical plane   145
    1.1   Technology and environment   145
    1.2   Industrial ecology and cleaner production   148
    1.3   Energy issues   151
2. Legal plane   160
    2.1   Environmental protection and sustainable
            development law   160
    2.2   Legal barriers to sustainable development   164
3. Political plane   166
    3.1   Policy and politics   166
    3.2   Democracy and sustainable development   171

CHAPTER 6: Integration of planes, the phenomenon of
globalization and the Sustainable Development Revolution   179

1. Overlapping of sustainable development planes   179
2. The challenges of globalization   181
3. Breakthroughs in human history   188
4. Sustainable development as a civilizational revolution   192

Conclusions   195

References   199

Index   227

Sustainable Development as a Civilizational Revolution – Pawłowski
© 2011 Taylor & Francis Group, London, ISBN 978-0-415-57860-8

About the Author

Artur Pawłowski, Ph.D., D.Sc. (habilitation), was born in 1969 in Chelm Lubelski, Poland.

In 1993 he received M.Sc. of the philosophy of nature and protection of the environment at the Catholic University of Lublin. Since that time he has been working in the Lublin University of Technology in the faculty of Environmental Protection Engineering.

In 1999 he defended Ph.D. thesis "Human's Responsibility for Nature" in the University of Card. Stefan Wyszynski in Warsaw.

Also at this University in 2009 he defended D.Sc. thesis "Sustainable Development—Idea, Philosophy and Practice".

Now he works on problems connected with multidimensional nature of sustainable development.

Member of International Association for Environmental Philosophy, Lublin Voivodship Board for Protection of Nature, and Polish Tourist Country-Lovers Association (PTTK).

Editor-in-chief of scientific journal "Problems of Sustainable Development".

Author of 95 publications (in English, Polish and Chinese).

Introduction

The concept of sustainable development has been formulated at a time when modern humanity gained the technological means to carry out almost any transformation of the world around us, but, at the same also got lost in the goals that their actions should serve. Our previous routes to development, based on the paradigms of the free market and economic growth, did not bring about the anticipated improvements for all mankind. Only few achieved prosperity, and this at the expense of environmental degradation and the increasing poverty of the majority of mankind. Moreover, previously unknown hazards emerged, such as the greenhouse effect or the ozone hole, and it seems no longer impossible that mankind could destroy the entire biosphere.

Subsequent civilizational challenges were answered with international programs and agreements. For many years, these mainly regarded the issues of natural conservation and environmental protection.

A breakthrough came in 1987, when the report "Our Common Future" was published by the UN, introducing the principle of sustainable development. This was described as a type of development which, while meeting present human needs, does so without threatening the ability of future generations to meet their own needs. This general statement became a starting point for the creation of specific action programs and was heavily publicized, particularly after The Earth Summit in Rio de Janeiro in 1992. When speaking of human needs, not only environmental issues were included, but also economical and social ones. However, the present author is of the opinion that this list should be further expanded in order to include philosophical issues (especially ethical issues related to human responsibility), as well as technical, legal and political issues.

This book aims at a critical analysis of the problems of sustainable development. It intends to demonstrate that the reflection previously conducted separately in natural, technical, social and philosophical sciences, may be coherent and mutually enriching. Furthermore, taking into account the fact that increasing attention is given to the problem within the UN, in the EU and also in the legislation of individual states, the work will postulate that implementing the idea of sustainable development may lead to an actual transformation of human relations with the both social and natural environment—so a significant turn in the history of mankind.

The author of this work is positive that implementing sustainable development will prove to be a revolution in human history, comparable to the earlier breakthroughs made when agriculture emerged and, later, with the development of science and technology. Let's make it happen!

Sustainable Development as a Civilizational Revolution – Pawłowski
© 2011 Taylor & Francis Group, London, ISBN 978-0-415-57860-8

CHAPTER 1

The evolution of the idea of sustainable development in history

The concept of sustainable development is a complex answer to the problems of the modern world. It significantly expands the scope of previous discussions concerning environmental protection. However, in order to show the full multi-dimensionality of the concept, its historical aspect must first be presented.

Recognition of the need to preserve nature occurred in the face of an increasing environmental degradation, resulting from the ever-expanding process of subordination of nature by humans. This process reaches far into the past even to the first appearance of humans on Earth. The earliest changes were purely local and caused, as far as we know, little or no disturbances in the environment. With territorial expansion, increase in the human population and its obtaining new skills—the scale of our impact on the environment also grew.

It is no mystery that many of the modern environmental problems date back far into the past. Smog, usually associated with the 20th century's environmental pollution, is one such example. Yet as early as 1542, the Spanish sailor Juan Rodrigez Cabrillo, observed a layer of fog as high as 300 m around Los Angeles, caused by the smoke from Indian campfires (Wojciechowski, 2001).

This chapter presents our road to sustainable development, which goes through various historic (mainly legal and political) initiatives, originally related to nature conservation and environmental protection, but later enhanced by other problem groups.

1 WORLDWIDE PERSPECTIVE

1.1 *Early initiatives*

Initiatives for protection of the environment have a long history. Even the primitive people occasionally took care of plants and animals that were especially useful to them. This care was quite radical; it included not only eradicating vermin, but competition species as well (Young, 1971).

Historical motives for protection of the environment very early included religious beliefs, associated with protecting places that were held sacred by the local communities. Other motives are presented in Table 1.

Probably the earliest formal decree on protection of the environment was introduced in China, during the reign of the Zhou Dynasty around 1122 B.C. It addressed the necessity to preserve the more valuable tree types, forests and green areas and to establish the office of forester. The decree was repeatedly reintroduced and found its place among the general rules of the forest economy,

Table 1. Motives for protection of the environment (Author's own work).

Motive	Short description
Biological, cultural and anthropocentric	The environment is crucial to human existence; hence it needs to be protected.
Economic	Damage to the environment corresponds to definite financial loss, which must be avoided.
Egoistic	Preserving the property of the ruler (e.g. the medieval regalia system), which in practice led to preserving nature as well.
Esthetic	Preserving the beauties of nature.
Ethical	Concerns the necessity of humans taking responsibility for nature.
Historical and patriotic (national)	Preserving locations of important events.
Ideal	Preserving nature for its own sake. This motivation is often associated with the ethical motive.
Religious	Preserving 'sacred places'.
Scientific	Answering the question: what action must be taken in order to preserve the natural environment?

which even included the financial support for afforestation of private property (Lisiecka et al., 1999).

Some rulers shared uncommon approaches to the environment. Among them was the Persian king Xerxes (519–464 BC). During an expedition through today's Turkey, near the town of Kallatebus, he came upon a beautiful plane-tree. In order to preserve the tree, a special sign was hung on it and a guard was left to see that the tree was not cut down (Lenkowa, 1981). This action can be seen to be associated with the ancient cult of trees. Their economic value was also recognized, so that there were both religious and economic motives for protection. Moreover, when wars broke out, the trees were often deliberately cut down by the conquerors.

Much attention was also paid to individual species. For instance in Europe, oak trees were cared for with special reverence. According to prehistoric beliefs, the oak was the first tree on Earth. Hell rested on its roots and its crown supported the Heavens. The Slavs surrounded the most magnificent trees with a fence with two wickets, through which only priests and princes were allowed to enter to pray.

This species had the status of a sacred tree and king of all plants in the ancient Greece as well. The rustle of its leaves was a guide for the priests in reading the divine judgments.

The same applies to Rome. Jupiter (lat. Iuppiter, identified with the Greek Zeus)—ruler of the gods, master of the Skies and of the Earth—wore oak wreaths and victorious commanders received crowns made of oak leaves. Oaks were important to common people as well. The dead were often buried underneath them. This tree was supposed to guarantee that no evil powers would disturb the dead.

Examples related to the role once assigned to oaks show that among the motives for protection of the environment (Bratkowski, 1991), those associated with the religious beliefs of that time were of great significance. Apart from trees, unexplained forces of nature were also worshipped, which entailed the protection of locations where religious ceremonies were frequently held. Among those sites, hills and wetlands were important. Legends also arose, which helped those areas to remain untouched.

Such species protection, as was introduced in ancient Asia, also had a religious character. For instance, as far back as the 3rd century B.C., during the reign of King Asoka, a decree was released concerning the protection of quite a wide range of inedible animals that were of no significance to humans, e.g. bats (Lenkowa, 1981). This was in accordance with the principles of Buddhism, which prohibited killing organisms, unless they were necessary to human survival (Auboyer, 1968).

The religious motive for protection was also important in later ages. It is worth pointing out the introduction of forest preservation near medieval Roman-Catholic monasteries. Forests were treated as places of contemplation and silent refuges, important to strengthen faith (Szafer, 1973).

A different—esthetic—approach to protection of the environment occurred in ancient Rome, China, Babylonia, Egypt and Greece. It used plant motifs in garden design (Boc et al., 2005). The perception of nature's esthetic values played a special role later in the age of romanticism. The beauty of nature untouched by the human hand was being compared to greatest pieces of art at the time.

Nature was also protected, because it constituted the ruler's properties, which could not be violated (egoistical motive for protection). Care for the ruler's property was the only reason for introducing such protection, but its effects had a much wider impact. In modern terms, we can say that nature was protected against its uncontrollable misuse. Even then it was recognised that resources would become depleted. Such a regulation was established in England (Canute I's Great "Charta de Foresta"—prohibiting deforestation and hunting in 1016 A.D.).

Introducing protection periods for fish (1030 in Scotland, 1258 in Spain, 1283 in England, in Poland during King Stephen Bathory's reign) was motivated differently. Economic issues were taken into consideration, expressing anxiety over the possible extermination of the most desirable species of fish, such as salmon. This not only included restrictions on fishing during spawning seasons, but also the prohibition of stunning fish or throwing poison to water. Breaking this law was severely punished, e.g. in England it was even punished by decapitation (Netboy, 1968). Sigismund of Luxemburg's decree was just as restrictive, prohibiting destruction of forests in the German Empire in 1436.

The Polish "Warta Statutes" of Wladyslaw Jagiello (Helcel, 1856) from the years 1420–1423, imposed hunting restrictions and forest protection. The latter included a postulate on the necessity to preserve rare and valuable tree species, especially yew trees, which were already being felled excessively. This was because yew wood was an ideal material for the production of bows and crossbows, which were the key weapons of the time. It is worth mentioning that the possibility of complete destruction of a species is still one of the main pillars of modern protection of the environment.

The issue of protecting individual species was continued in Poland, e.g. the wisent (European bison). The first warnings of its possible extinction date back to the 16th century. Already in 1541 its refuge in the famous Bialowieza Forest was taken into royal care, and the wisent itself was considered a royal animal, whose hunting was prohibited. This was strongly emphasized in "The Forest Charter" of 1557 (Radecki, 1989).

More detailed regulations can be found in the "Statutes of Lithuania" (declared in 1529, 1566 and 1588). These included the issue of forest protection and introduced species protection for wild animals, especially beavers and the wisent already mentioned. Even the landowner was not allowed to carry out any work in the vicinity of a beaver's dam. Hence, this was not a simple hunting restriction, but a complex protection of the beaver's biotope! Even the case of when a beaver leaves its dam and creates a new one elsewhere, was taken into consideration!

Another type of motivation was evident in 1535 in the canton of Zurich, Switzerland, where protection of birds was introduced based on the beauty of their singing voices (Lenkowa, 1981). An esthetic motive was not predominant in this case, since the useful role of those birds was stressed, namely, hunting pests (mainly insects) in forests and rural areas.

Solutions including wider environmental conditions were applied in the 16th century, when the first nature reserve was established in 1576 in the forests of the Hague region—see Table 2. Several others were established e.g. in Schleswig-Holstein in 1671. The forests there were even described as the greatest magnificence given to the princedom by God (Lenkowa, 1981), and in 1713 Hans C. von Carlowitz began even the discussion on sustainable forestry—see Table 6 in chapter 2.

More reserves were set up in Europe in the 19th century. This resulted from the changes, which occurred at the turn of the 18th and 19th century, related to

Table 2. Europe's first natural reserves (Lenkowa, 1981; Michajlow, 1978).

Year	Site of reserve and its character
1576	Hague region, forest reserve.
1668	Baumann's Cave in the Harz Mountains.
1671	Schleswig-Holstein, forest conservation.
1703	Izmailovsky forest near Moscow.
1765	Monastery forest, Dnieper river.
1803	Theresa's forest near Bamberg (Bavaria).
1805	Gammelmosen peatbog in Denmark.
1824	Luisa's forest near Bamberg (Bavaria).
1836	Dragon's Rock near Bonn.
1852	Devil's Wall built of sandstone, near Thale and Blankenburg (Tyrol).
1838	Forests near the town of Nové Hrady.
1838	Hojna Voda forest in Nové Hrady Mountains.
1844	Conservation of the peatbogs near Copenhagen.
1853	Fontainebleau forest near Paris.
1858	Forest on Mt. Boubin near Šumava.
1877	Moors in the Sempt river valley near the town of Landshut (Bavaria).
1888	Plowed steppes in Askania-Nova near the Dnieper river. Several other reserves were founded soon, including Ukraine's steppes in Volhynia and in the Voronezh region.

excessive exploitation of subtropical areas, colonized by the empires of the time. For instance, already in 1560 a significant deforestation was observed in the West Indies. Recognition of the problem and the attempts at solving it were associated with the presence of scientists in the expeditions (Grove, 1992). Thanks to them, an innovative law was introduced to the British colonies in North America in 1681, introducing wide forest conservation; subsequently a decree was issued, ordering that every fifth acre was to be left intact during deforestation.

In 1764, the first rainforest reserves were set up on Tobago Island, initially covering 20% of its territory (Grove, 1992).

The island of Mauritius is another significant example. At first it was under Portuguese rule, then Dutch and since 1721, French. It was scientists of the latter nation, led by Philibert Commerson and Bernardin de Saint Pierre, who noticed the large-scale devastation of the island's forests, especially in the more accessible coastal areas. The account included important words: "the balance between man and nature was disturbed on Mauritius" (Grove, 1992). Fortunately, in 1769, with support from the Governor of Mauritius— Jesuit Pierre Poivra, the island was taken under legal protection, motivated by nature's value for its own sake as well as by the negative effects, to the local climate, of cutting down forests. Legal regulations were radicalized in 1803, when deforestation of mountain slopes (above one third of their height) was completely prohibited.

In 1852, the Scottish scientists: Alexander Gibson, Edward Balfour and Hugh F.C. Cleghorn published their report on the catastrophic deforestation taking place in India (Grove, 1992). It included an innovative warning that lack of preventive action against further degradation in the region would not only lead to the destruction of nature, but also to negative social effects. Among others, the possibility of droughts resulting from reduced rainfall which may result in food shortages, was pointed out. In the face of repeated climate disturbance (the first droughts has already occurred during the deforestation periods, in 1835 and 1839) and of the specter of hunger, appropriate legal actions were initiated. In 1864, even a special forestry unit was established, whose task was the policing of legal regulations in that area.

The issue of forest conservation was also raised in Poland at the turn of the 17th and 18th century. In 1778, King Stanislaus August Poniatowski passed "The Forest Proclamation" (Radecki, 1989). This introduced an explicit prohibition of uncontrolled deforestation. The threat of complete deforestation in the entire country was also emphasized. That is a motive for preserving forests as part of national heritage! Another innovation was that not only was the document announced and printed, but its regulations were also made public in parishes. This makes an educational postulate, which—from today's point of view—would fit in environmental education programs.

In 1863, "The Alcali Act", the first legal document, concerning the reduction of environmental pollution, was signed in Great Britain (Mullerscience. com, 2009).

During the same period, some very interesting solutions were adopted in Galicia by the National Parliament in Lvov, on the initiative of the Polish members (signed afterwards by the Emperor in Vienna). Two acts deserve special distinction.

In 1868, the Parliament passed the act (signed by the Emperor a year later) "On the Prohibition of Capturing, Eradicating and Selling Wild Alpine

Animals, Characteristic for the Tatras, the Marmot and Chamois" (Boc et al., 2005). Fines were set for violating the prohibition and in case of inability to pay the fine—a detention penalty was imposed. Such action may be defined as direct species preservation. What is particularly significant is that this was the first act in Poland to be inspired by scientific research. Namely, in this case it was the works of a scientist from the Physiographic Committee, established in 1865 by the Cracow Scientific Society (Radecki, 1989).

Also in 1868, the National Parliament in Lvov passed the "Act on Prohibition of Capturing and Selling of Singing and Insectivorous Birds" (Radecki, 1989). The document had to wait for the imperial signature for ca. six years, till 1874. During that time it had been modified and its name changed to the "Act on Preserving Some Animals Useful for Agriculture" (Boc et al., 2005). The document prohibited removing or destroying eggs and nests of all wild yet harmless birds, as well as catching and killing birds. A single exception was made for scientific purposes. Moreover, the act includes a precise list of the protected birds, the penalties for non-compliance and assigned offices responsible for law enforcement. The structure of the document was not vastly different from that of today's legal acts. Further, it offered solutions that even today would be considered innovative. Among the adopted regulations was one that obliged teachers in regular and Sunday schools, to teach their pupils of the harmfulness of taking out nests, catching and killing useful birds, and remind them of the provisions of the act every year before the breeding season (Boc et al., 2005). Therefore, this was—using today's language—an obligatory environmental education! And one that not only specified its contents, but also how and when they were to be taught.

In 1872, the world's first national park was founded in Yellowstone, positioned on the borderline of three American states: Wyoming, Montana and Idaho. This was the result of political pressure from a group of enthusiasts led by Ferdinand Vadiveer Hayden (Yellowstone, 2007). This group first managed to convince the Congress to fund a scientific expedition, which culminated in a 500-pages long documentation of the region's nature. This later provided the basis for the creation of the park. Other American parks, Yosemite and Sequoia, were founded in 1890 and in 1899 Mt. Rainer.

The first national parks in Europe (Walczak et al., 2001) were founded at the beginning of the 20th century. These were Abisko, Sarek in Sweden (1909) and Suisse in Switzerland (1914).

During that period, other initiatives were also taken for nature preservation; some of them were even international. It is worth mentioning the "Act on the Protection of Birds" in Great Britain in 1868 (nearly a decade earlier, in 1860, the British introduced bird protection in one of their colonies—in Tasmania), or the agreement of 1883 signed by Germany, Netherlands and Switzerland, regarding salmon protection in the Rhine basin (Lenkowa, 1981; Grove, 1992).

However, success was not always achieved. This was the case with an initiative of the Swedish government in 1872, concerning the foundation of an international committee on the protection of migrating birds in Europe (Eckerberg, 1997). Also, "The Convention on Whale Hunting Restrictions" of 1931, although signed by 24 countries, was not regarded as successful. First of all, it placed only few restrictions on its signatories and secondly, the two countries that hunted the whales most intensively (USSR and Japan), refused to sign the document.

1.2 *Contemporary later advances*

The beginnings of serious activities towards protection of the environment in Europe date back to the turn of the 19th and 20th century. Hugo Conwentz's (1855–1922) movement towards the protection of natural monuments (called Naturdenkmalpflege) is a symbol of this. Although the notion of a natural monument was already introduced in 1819 by the German geographer, scientist and traveler, Alexander von Humboldt, it was Conwentz who managed to give it publicity and perform actual protective activity. The movement gained followers in many countries.

At the beginning of the 20th century, the first acts on protection of the environment were passed: in 1902 in Germany, in 1906 in France and in 1910 in Norway (Walczak et al., 2001).

A conference organized by the Swiss government in Bern on 17 Nov 1913 was also an important initiative. It gathered representatives from seventeen countries, who established the International Advisory Board for Protection of Nature (Commission Consultative pour la Protection Internationale de la Nature), headquartered in Basel (Lenkowa, 1981). The organization constituted in 1914, but its further activity was disrupted by the outbreak of World War I.

A similar initiative was presented in 1928 during the 5th Congress of the International Biology Union. A year later, the International Union for Conservation of Nature (Bureau International pour la Protection de la Nature), headquartered in Brussels, was established (Lenkowa, 1981). At first, it operated as a private institution, obtaining proper legal personality in 1934. Its activity was ended by the outbreak of World War II.

During the interwar period, national acts concerning nature conservation were also significant, e.g. in 1930 such an act was passed in France and in 1934 in Poland.

Later on, the establishment of the United Nations (UN) and with it the UNESCO (United Nations Educational, Scientific and Cultural Organization), was an important moment (see Table 3)—and not only for the historic aspect of nature conservation. Among its founders were the representatives of only 50 countries, today the UN includes 192 countries.

Despite many problems, resulting from the devastation laid by the World War II, environmental issues were taken up rapidly by the UN. "The Universal Declaration of Human Rights" passed in 1948, became a reference point (Unic.org, 2009). It confirmed that everyone is free (Article 1), has the right to live (Article 3), to work (Article 23), to education (Article 26), to a standard of living adequate for the health and well-being of himself and his family (Article 25) and is equal before the law (Article 7) regardless of their sex, race or religion (Article 2).

Also in 1946, a UNESCO-based International Union for Conservation of Nature (IUCN, originally: International Union for the Protection of Nature) was appointed. At present, it associates 1000 organizations from 160 countries.

From the very beginning, the Union aimed not only at creating and registering new sensitive areas, but concern over an increasing range of global threats to ecology was also expressed. This was a major step forward. While problems

Table 3. Selected modern international initiatives for protection of the environment and sustainable development (Author's own work).

Year	Initiative
1945	Creation of the UN and UNESCO. Among the first UN initiatives was the creation of the Economic and Social Council (ECOSOC).
1946	Creation of the IUCN—International Union for Conservation of Nature.
1948	Passing of the "Universal Declaration of Human Rights".
1960	The first ten-year development strategy prepared by the UN (10-year International Development Strategies).
1962	Publication of the book "Silent Spring" by Rachel Carson. This was the first celebrated book pointing out the threats associated with the use of pesticides, and therefore, at the issue of an advancing degradation of the environment.
1966	Launching of UNDP—United Nations Development Programme.
1969	U'Thant's report "The Problems of Human Environment" at the UN on negative consequences of environmental degradation.
1969	Creation of SCOPE—Scientific Committee on Problems of the Environment.
1970	Creation of the UNESCO Man and the Biosphere (MaB) program.
1972	Stockholm Conference and the Stockholm Declaration. The first report from the Club of Rome: subsequent reports still appear today.
1972	Creation of the UNEP—United Nations Environmental Programme.
1974	Establishment of Worldwatch Institute. This independent American organization prepares reports on the state of the world to this day.
1978	Establishment of the UN Habitat program, concerning the issue of rapid urbanization.
1980	Development and publication of "Nature Conservation Strategy" for IUCN.
1982	UN "World Charter for Nature".
1983	Appointment of WCED—World Commission on Environment and Development.
1987	Report "Our Common Future" developed by the WCED. This publication introduced the notion of sustainable development to the UN documents.
1990	Establishment of ICEI—International Council for Environmental Initiatives, reshaped later into a council dealing with the introduction of sustainable development at regional and local levels.
1992	Earth Summit in Rio de Janeiro, expansion of the concept of sustainable development: "Rio Declaration", "Convention on Biological Diversity", "Convention on Climate Change", "Forest Principles". Moreover, a new strategy for action—"Agenda 21"—was prepared. In order to watch over its realization, the Division for Sustainable Development (DSD) was appointed, which operated within the Department of Economic and Social Affairs (DESA).
2000	Passing of the "Millennium Declaration", regarding the report "We the Peoples—the Role of the United Nations in the 21st century", prepared by the Secretary-General K.A. Annan.
2000	"The Earth Charter".
2002	Earth Summit in Johannesburg, supporting the legitimacy of developing the concept of sustainable development.

of protection were so far usually limited to nature and conservation of the most valuable areas, the documents of the Union pointed at equally important issues related to degradation of the whole environment. However, these postulates were insufficiently publicized by the media, which limited the Union's activity in the field of environmental protection.

Work covering protection of the environment were expanded in 1961 along with the creation (under the aegis of the UN) of the World Wide Fund for Nature (WWF).

Among other early UN institutions, the following should be distinguished:

- Food and Agriculture Organization (FAO), established in 1945, whose goal was to solve problems concerning the shortage of food in impoverished regions of the world.

  In 1995, a special unit was formed within this organization (Sustainable Development Department). The goal remained unchanged, but the scope of factors included was expanded to include problem groups on sustainable development, thus not only covering food issues, but also the full biophysical and socio-economic contexts.
- World Health Organization (WHO), established in 1948. Since 1963, it has been carrying out specific aid programs: The World Food Programmes (WFPs). At present, the definition in force within this organization goes beyond the problems of health and medical conditions and includes physical health as well as mental and social conditions of human development.
- United Nations Industrial Development Organization (UNIDO), established in 1966. It also touches upon issues of industry's impact on the environment.
- United Nations Development Programme (UNDP), formulated in 1966, which is more general than UNIDO. Its main goal is to reduce the level of poverty in the whole world. In 1993, a new program was implemented within the UNDP—CAPACITY 21—whose main task is to aid individual countries in implementing the strategies of sustainable development.

It has to be stressed that each and every one of these activities originally concerned a specific, narrow group of problems, expanding its horizons with time by other aspects.

In 1960, the first of the 10-year International Development Strategies was announced within the UN (UN, 2009). This initiated a plan to achieve a higher standard of living, full employment, economic and social progress. These are also major goals in the era of sustainable development.

A year later, the Organization for Economic Co-operation and Development (OECD) was created. This was not a UN initiative, although it united the rich countries within the concept of increasing efficiency and promoting free markets. Further, the program included aiding poorer countries in their own development (Gupta, 2002).

Also outside the UN, two important publications appeared.

The first of them was "Silent Spring" by Rachel Carson (Carson, 1962). It was a warning against further use of chemicals in the environment (as the author points out, around 500 new chemical compounds are introduced to the environment every year), especially pesticides, such as DDT. The title of the book anticipates the situation when birds, made extinct by pesticide poisoning, will no longer be heard singing the following spring. The publication found great interest and was one of the major causes of prohibiting the use of DDT in the United States in 1972. Although the pesticide was synthesized for reducing vermin populations, especially insects, it turned out to be harmful to the whole biocoenosis (especially predators).

Carson's book undermined the faith in unlimited human capability to control the environment with the help of science and technology. The author warned, that science armed itself with the latest, dreadful types of weapons; and aiming them at insects, it aimed them at the Earth (Kraoll, 2006). One of the effects of this publication was the creation of the Silent Spring Institute, still operating today, which deals with environmental and health issues alike (Silentspring.org, 2011).

A second important publication, in 1969, was "Subversive Science" (Shephard & McKiney, 1969). One of its authors, ecologist and philosopher Paul Shephard, pointed out in the introduction, the necessity to reclaim the disturbed balance in the modern world. This task is interdisciplinary, with particular emphasis on ecology, due to the holistic perspective rooted in its essence (Shephard, 1969).

In 1969, two further important events occurred:

- From a regional perspective one was the passing of the "National Environmental Policy Act" (NEPA) in the USA, which came into force on 1 January 1970.
- From global viewpoint, the famous report of the UN Secretary-General U'Thant, "The Problems of Human Environment". During the succeeding decade it was the most quoted document in the whole history of the United Nations.

The NEPA stressed the necessity to adopt an interdisciplinary approach in decision-making processes concerning environmental issues, while using natural sciences, as well as the planning and managing of protection of the environment. This document was precursory, since not only the necessity of caring for the environment was highlighted, but it was also recognized that effective action would only be possible with an integrated approach going beyond traditional environmental protection.

In contrast, U'Thant's address was the result of the discussion started at the UN a year earlier when, during the 23rd UN General Assembly a resolution on environmental problems was passed, which obliged the Secretary-General to prepare such a report. The report "The Problems of Human Environment" was delivered on 26 May 1969, during the 24th session of the UN General Assembly. It stated that, for the first time in human history, a global crisis occurred, including both the developed and the developing countries—a crisis regarding human approach toward the environment. Its signs had been visible for a long time—demographic explosion, insufficient integration of over-developed technology with the requirements of the environment, destruction of cultivated areas, unplanned urban development, reduction of free areas and the increasing threat of extinction of many animal and plant life forms (Tobera, 1988). The conclusion was also important—we all live in the same biosphere, whose space and resources, however enormous they may appear, are limited.

U'Thant did not restrict himself to discussing only major environmental threats. He also argued that seeking alternative routes for human progress will do no good, unless biological and social aspects are analyzed separately from the issue of physical degradation of the environment. Those were undoubtedly the pillars for the formulation of the sustainable development concept!

A significant consequence of that report was the attempt at a wider international cooperation within the UN to protect the environment. Moreover, the general principles of the report were publicized by the media, which had a major impact on their popularity, as well as on social support for particular programs.

Subsequent UN initiatives are characterized by variations in the scope of the issues in question. Some of the proposals were quite close to interdisciplinary characteristic of sustainable development; others only included some of its aspects.

On 30 July 1969, the UN General Assembly had passed the declaration "On Social Progress and Development". The issue of further successful human development was clearly combined with the need to protect nature (Papuzinski, 1999), and the discussion was placed in the social context.

In the same year, the Scientific Committee on Problems of the Environment (SCOPE) was appointed. This organization still prepares reports on the global environmental issues.

In 1970, a UNESCO program Man and the Biosphere (MaB) was created. In the relation between human and the environment, not only ecological issues (most clearly relating to the World Biosphere Reserves, established within this project) were touched upon, but also social and economic issues.

Another major step, which is probably the most significant effect of U'Thant's report, is the so-called 'Earth Summit', or 'Stockholm Conference', which took place on 5–16 June 1972.

The meeting was preceded by the publication of the report "Only One World" (Ryden et al., 2003), concerning the status of the environment, along with 200 detailed documents, prepared by both the UN and by the governments of individual countries (including the Holy See), as well as by scientific and social organizations. 130 delegations took part in the conference and delegates were tasked within three working committees (Ryden et al., 2003):

- Social and cultural aspects of protection of the environment.
- Natural resources (mainly the issue of their exhaustibility).
- International aspects of the struggle against environmental degradation (resources availability, actions taken and the appointed organizations).

It should be stressed that the discussion included the cultural limitations of the world and therefore went beyond the basic problems both at the ecologic and the social level.

Among the documents signed in Stockholm, the Stockholm Declaration ("Declaration of the United Nations on the Human Environment") played an important role (Ryden et al., 2003). It consists of two parts:

The first part refers to the goals and tasks of protection of the environment in the global perspective. It has been stated that a point has been reached, which—due to the rapid progress in science and technology—allows for reshaping the humans' natural environment on an unprecedented scale. Whereas both the environment created by nature and the one created by humans are necessary to our survival, wisely used, the human ability to reshape the environment may provide benefits to all nations, as well as give them the opportunity to improve the quality of life. The very same ability—misused or used unilaterally—may cause immense damage to the humans and the environment (UNEP, 2009). Among the global developmental challenges, the necessity to maintain peace was pointed out, as well as the issues of socio-economic development (clearly stressing the problems of the developing countries and also calling on the rich countries for help). The road to protection of the environment and actions for improving the state of the environment were pointed as proposals for the future. Its goal was set, not only in the rights of present generations, but of the future generations as well, which is an important point on the road to formulating the concept of sustainable development.

The second part of the Stockholm Declaration is a set of 26 principles, addressed to the governments of individual countries, as well as local authorities. Principle 13 stresses the necessity to adopt an integrated and coordinated development plan, providing compliance between the development and the need to preserve the environment for citizens. Principle 21 is also worth mentioning: it states that, according to the "United Nations Charter" and the principles of international law, countries have the sovereign right to exploit their resources in compliance with their environmental policies and have the obligation to ensure that the activity within their supervision causes no harm to the environment of other countries (Bergström, 1992). This was, therefore, an attempt at a compromise between the previous expansive model of civilization development and acceptable restrictions, which would reduce human pressure on the environment.

The Stockholm Conference was an important event in the history of the UN. However, regardless of the proposed official solutions, it also showed how—typical for that time—political animosities between East and West posed a significant barrier at the international level. As it turned out, despite the invitation, delegations from the communist countries did not arrive. This was not caused by the issues taken in Stockholm, nor was it any form of resistance to protection of the environment. The decision was purely political and was a protest against the non-recognition of East Germany (at the time part of the Eastern bloc of countries controlled by the USSR) by the western countries. Fortunately, even in the East the deliberations were diligently observed. It is no coincidence, that four years later two new pro-environmental regulations were introduced to the Polish Constitution, compliant with the spirit of the Stockholm Declaration.

The Stockholm Conference entailed other UN initiatives.

At the regional level, it was "The Stockholm Convention" in 1974, concerning the conservation of the Baltic Sea, and appointing the Helsinki Commission (HELCOM) to watch over the realization of the goals. This activity was continued, and an updated version of the convention was passed in 1992 (Ryden et al., 2003).

At the global level, in 1972, the UNESCO General Conference in Paris passed the "Convention on the Protection of World Cultural and Natural Heritage" (Ryden et al., 2003). The starting point was noticing the threats, which not only regarded nature, but also the objects of culture, and which are the result of modern social and economic transformations. It has to be emphasized, that the world of nature was treated here as equal to the world of culture. The convention placed responsibility for maintenance of the heritage on individual countries, and a specific instrument was the list of the most precious areas and monuments in the world—the World Heritage (WH) List.

Among other initiatives, an important role was played by the Resolution of UN General Assembly, passed on 15 January 1974 "On Co-operation in the Field of Environmental Protection Regarding Natural Resources Belonging to Two or More Countries" (Boc et al., 2005). It was concerned with the issue of transboundary transfer of pollution, which was a major issue in some border regions in Europe.

Another initiative, which was a direct result of the Stockholm Conference, was the appointment of the United Nations Environmental Programme (UNEP) in 1972. Its main goals were:

- To monitor the condition of the environment.
- To support research on new scientific and technological solutions.
- To develop new strategies (referred to as 'Action Plans').
- To initiate the development and implementation of international agreements on environmental protection.

It is estimated that UNEP's engagement contributed to negotiating as much as two thirds of international environmental treaties (Kozlowski, 2005).

Also within the UNEP, in 1973, a concept of eco-development was introduced, concerning three levels (Kozlowski, 2005):

- Estimating the cost of human impact on the environment, and taking into account social costs.
- Environmental management.
- Environmental policy.

The discussion was modified to include more details in 1975, when—during the 2nd session of the Programme Governing Council—an important postulate was passed, to guarantee such a course of inevitable economic development, that would not disturb the human environment irreversibly, one that would not lead to degradation of the biosphere and would reconcile the laws of nature, economy and culture alike (Timoshenko & Berman, 1996). This description is very reminiscent of modern definitions of sustainable development.

However, the beginning of the 1970's was marked not only by UN initiatives.

The establishment of the world's first 'green' party in 1972 deserves special note. It was the Values Party, founded in New Zealand (Greens.org.nz, 2008).

The first publications from the Club of Rome were also in the 70's. This association still exists today (Clubofrome.org, 2010) and brings together an international group of entrepreneurs, statesmen, and scientists, among whom the leading positions are occupied by scientists from the Massachusetts Institute of Technology (MIT) led by Denis L. Meadows.

The initiator for the establishment of the Club was A. Peccei, who had organized the first meeting in Rome in 1968. The Club was registered in Geneva in 1973; however, its first report was already published in 1972. The publication, titled "Limits to Growth" (Meadows et al., 1972), placed the discussion in an economic context and confirmed U'Thant's thesis, that there are limits in nature, exceeding which (be it by overexploitation of the natural resources or by an excessive increase in pollution) may lead to a collapse of balance in the biosphere. It is worth mentioning that a similar vision was also presented by Jay W. Forrester (Forrester, 1971).

The estimate performed by the Club of Rome, based on specially prepared computer model called World 3, referred to the global perspective. It has been pointed out, that if the present trends—marked by the exponential increase of pollution of the environment—do not change, the anticipated catastrophe will occur within the next 100 years. However, there are still opportunities to change the course of events. Therefore, the conclusion to "Limits to Growth" includes a proposal of an alternative road of development, leading to a state of global balance, within which the basic material needs of every human being on Earth would be satisfied, and everyone would have the opportunity to make use of their capabilities (Meadows et al., 1972). This reasoning is very close to that of sustainable development: it contains clear reference to the quality of human life and to the conditions determining it now and in the future.

While making the balance of profit and loss, the report proposed—as a solution guaranteeing balance and a secure future—the concept of 'zero growth' which imposed restrictions at the demographic level (the issue of population growth) and the environmental level (reducing the consumption of natural resources, especially non-renewables). 'Zero growth' would then mean balancing the birth rate and the number of deaths on a global scale, as well as developing such forms of human activity as education and scientific research not related to industry and consuming no resources.

The discussion was expanded in the next report from the Club of Rome, "Mankind at the Turning Point" (Mersarovic & Pestel, 1975). It suggested that modern environmental crises (i.e. energy, food, resource crisis) are not temporary, but are a lasting result of the historically dominant trends in development. Solving these problems is only possible at the level of global cooperation. As the authors state: we need a full integration of all layers of our hierarchical model of the world, i.e. to simultaneously consider all aspects of human evolution, from individual systems of values, to ecologic and environmental conditions (Mersarovic & Pestel, 1975). Such an interdisciplinary approach is compliant with the concept of sustainable development.

In the practical aspect, the report "Mankind at the Turning Point" instead of the radical 'zero growth' offers a new idea of 'limited growth'. It has also been pointed out that even now the development of particular areas of Earth runs at different speeds; the impact of the population on the environment is also variable. Reducing the differences would lead to more fair world, whereas closer cooperation would enable a more rational use of natural resources. The issue of opposition between the rich and the poor countries was also the subject matter of the following two reports.

In 1976 a new study was released, "Rio Report: Reshaping the International Order" (Tinbergen, 1976). Here it was argued that the main cause of the world's developmental problems was the inequity of international systems, and among the major obstacles on the way towards improvement, the arms race was highlighted. It absorbs enormous funds, and the weapons themselves are a threat to the environment on a global scale.

The report "Goals for Mankind" (Laszlo, 1977) also contributes to this school of thought. Similar to modern studies regarding sustainable development, it adopted three main perspectives: international, regional and local. On these was based the attempt on determining global goals. Within the last group, the following were pointed out: the necessity of ensuring global security and maintaining peace, environmental issues (power and resources), moreover, once again attempts were made at improving the condition of poor countries (i.e. considering. opportunities of socio-economic advancement for the people living in those countries).

Out of the obstacles, the 'inner limits' were distinguished. These refer to the decision levels of the rich countries, their tendency to give priority almost completely to their own benefits and their reluctance to share their wealth. Breaking down 'inner limits' was even described as a moral imperative, which needs to be stressed, since ethical statements in such reports are not common.

Subsequent reports from the Club of Rome appeared regularly and were proposed as answers to contemporary changes in the global situation[1]. Despite extensive discussion they induced, the issues outlined in them have not been solved to this day, mainly due to the 'inner limits' mentioned above.

Another important independent initiative, which dates back to the 70's, was the appointment of independent Worldwatch Institute in 1974 (Worldwatch. org, 2010). Of the many publications from this American organization, a special position is occupied by the yearbook "The State of the World". Its first edition was released in 1975. In this case also, environmental issues, dominant in the first studies, were later enhanced by a wider context (those reports are discussed in the final part of this chapter).

In 1978 the UN Habitat program was established (Unhabitat.org, 2010), devoted to issues of rapid urbanization, especially in the Third World countries. The program was promoted with the slogan 'Shelter for All'. It presented environmental issues (i.e. lack of housing, lack of drinking water), pointing, at the same time, at the wider social, political and economic issues.

---

[1] According to the official Club of Rome list, the full set of reports includes: D.H. Meadows, D.L. Meadows & W.W. Behrens, "The Limits to Growth" (1972); M. Mersarovic & E. Pestel, "Mankind at the Turning Point" (1974); J. Tinbergen, "Rio Report: Reshaping the International Order" (1976); D. Gabor, "Beyond the Age of Waste" (1978); E. Laszlo, "Goals for Mankind" (1977); T. de Montbrial, "Energy: the Countdown" (1978); J. Botkin, M. Elmandrja & M. Malitza, "No Limits to Learning" (1978); M. Gauernier & Tiers-Monde, "Trois Quart Du Monde" (1980); O. Giarini, "Dialogue on Wealth and Welfare, an Alternative View of World Capital Formation" (1980); R. Hawrylyshyn, Road "Maps to the Future, Towards More Effective Societies" (1980); J. Saint-Geours, "L'Imperatif de Cooperation Mord-Sud, La Synergie Des Mondes" (1981); A. Schaff & G. Friedrichs, "Microelectronics and Society: for Better and for Worse" (1982); E. Mann Borgese, "The Future of the Oceans" (1986); R. Lenoir, "Le Tiers Monde Peut se Nuourrir" (1984); B. Schneider, "The Barefoot Revolution" (1988); E. Pestel, "Beyond the Limits to Growth" (1989), O. Giarini & W.R. Stahel, "The Limits to Certainty" (1989/1993); A. Lemma & P. Malaska, "Africa Beyond Famine" (1989); A. King & B. Schneider, "The First Global Revolution" (1991); D.H. Meadows, D.L. Meadows & J. Randers, "Beyond the Limits: Confronting Global Collapse, Envisioning a Sustainable Future" (1993); Y. Dror, "The Capacity to Govern" (1994); B. Schneider, "The Scandal and the Shame: Poverty and Underdevelopment" (1995); W. van Dieren (ed.), "Taking Nature into Account, Towards a Sustainable National Income" (1995); E.U. Von Weizsäcker, A.B. Lovins & L.H. Lovins, "Factor Four, Doubling Wealth—Halving Resource Use" (1997); P.L. Berger, "The Limits of Social Cohesion: Conflict and Understanding in a Pluralistic Society" (1997): O. Giarini & P. Liedtke, "Wie Wir Arbeiten Werden" (1998); E. Mann Borgese, "The Oceanic Circle: Governing The Seas as a Global Resource" (1998); J.L. Cebrian, "In Netz: Die Hypnotiserte Gesellschaft" (1999); R. Mohn, "Menschlichkeit Gewinnt" (2000); S.P. Kapitza, "Information Society and the Demographic Revolution" (2001); F. Vester, "Die Kunst Vernetzt Zu Denken" (2002); O. Giasrini & M. Malitza, "The Double Helix of Learning and Work" (2003); D. Meadows, "Limits to Growth—the 30-year Update" (2004); E.U. von Weizsäcker, "Limits to Privatization—How to Avoid Too Much of a Good Thing" (2005); E.U. von Weizsäcker, K. Hargroves, M.H. Smith, C. Desha & P. Srasinopoulos, "Factor Five, Transforming the Global Economy through 80% Improvements in Resource Productivity" (2009).

15

A year later, on 13 November 1979 (the document entered into force in 1983), one of the most significant international agreements was signed in Geneva: "Convention on Long-Range Transboundary Air Pollution". Article 1 states that the case is made with air pollution, whose physical source is wholly or partly located within the jurisdiction of one country, and whose negative effects have an impact on the territory within the jurisdiction of another country, over such distance, which makes it impossible to distinguish the share of individual sources—or groups of sources—to the total emission. An important addition to this was "The Madrid Convention on Transfrontier Co-operation between Territorial Communities or Authorities", passed in 1980.

At that time the issue was clearly exemplified in Europe, by the so-called 'black triangle' at the confluence of the Polish, East German and Czechoslovakian borders. Enormous coal power plants were localized in that region and—because of the prevailing winds—Polish spruce forests in Karkonosze Mountains and Izerskie Mountains were dramatically affected (estimates show that the whole region was the source of ca. 30% of the total emission of sulfur oxides in Europe). After 1989 however, new technologies were introduced and pollution levels were significantly decreased within a decade.

An important events in 1980 was the foundation of the Polish Ecological Club (PKE) in Cracow, Poland (Runc, 1998). This was the first independent, non-governmental organization of this type within the circle of communist states.

The Club's ideological declaration contained some important phrases (Juchnowicz, 2006):

- The Polish Ecological Club is a social movement of people aware of the threats following the biological imbalance brought about by technological civilization and a consumer model of life, working for the good of the nation, for the protection of nature and of human environment.
- Humans have the fundamental right for freedom, equality, decent living conditions in an environment, whose quality should allow them to sustain their dignity and well-being.

Also in 1980, the IUCN and the UNEP released the "World Conservation Strategy". Detailed goals for the strategy can be summarized as follows (IUCN, 1980):

1. To maintain the fundamental ecological processes and systems, which are refuges for life. Therefore, conservation of soils, green areas, forests are included here, as well as such processes as e.g. the self-purification of water.
2. To preserve genetic diversity (which later came to be termed 'biodiversity').
3. To ensure sustainable development (usage) of land and ecosystems.

The document's general purpose was to integrate the conditions for protection of the environment and development, and thus ensure an optimal habitat for all humans. This almost all-encompassing goal goes far beyond solely environmental issues. At the very beginning of the document, significant phrases are included: Human beings, in their drive to achieve economic development and exploit natural resources, must accept the fact that the resources, as well as the ecosystem's capacity, are limited and must take into consideration the needs of future generations (IUCN, 1980). This reasoning is very close to the principle of sustainable development. What is more, in the strategy's subtitle (and in point 3 of its main text), we can even find the phrase 'sustainable development', although this was not the full formulation of the concept.

"The World Conservation Strategy" was further elaborated on 28 October 1982, when the UN General Assembly passed the document entitled "World Charter for Nature" (UN, 1982). It clearly revealed our changed approach to the environment, which occurred in the latter half of the 20th century. It is worth mentioning, that the document was based on the UN "Universal Declaration of Human Rights" of 1948 and called for respect for the nature. As stated in the early part of the introduction, humans and their civilization are fixed in nature and it was the latter, which enabled numerous human achievements, both artistic and scientific. Moreover, living in harmony with nature ensures the best development for mankind.

It is worth quoting another fragment of the introduction to the Charter which states that Man must acquire the knowledge to maintain and enhance his ability to use natural resources in a manner which ensures the preservation of the species and ecosystems for the benefit of present and future generations (UN, 1982). Although the notion of sustainable development was not mentioned specifically, the above formulation is entirely consistent with the principle of such development. Moreover, a number of issues were pointed out, which also became the main focal points of sustainable development (UN, 1982):

- Preserving biological diversity.
- The necessity that proper socio-economic development must include the issues of protection of the environment.
- Suggesting a long-range assessment of the actions, especially the necessity to assess the effects of such actions, which might contribute to the degradation of the environment.
- The interconnection between the issue of population growth and increasing the living standard, and that of natural systems' capacity.
- Pointing out the threats related to conflicts.
- The necessity of reducing the consumption of non-renewable resources.

The wide range of problems shown in the "World Charter for Nature" was discussed during the special session of the UNEP—which took place in Nairobi in 1982 (Swierczek, 1990). Problems with implementing strategies adopted since the Stockholm Conference were underlined, and the barriers between the rich and the poor countries were presented as their main cause.

This issue was also discussed by the independent World Commission on Environment and Development (WCED), established in 1983. The Prime Minister of Norway, Gro Harlem Brundtland, was appointed head of the Commission. Using all the UN's resources, the report "Our Common Future" was prepared in 1987. It was an attempt at a holistic approach to the problems of the modern world. A common, narrow understanding of the notion of 'development' (only including purely economic development) was warned against, as well as the equally narrow approach to the notion of 'environment'. In the modern world—as clearly stressed in the commentary by Donald J. Johnston on behalf of OECD—the environment does not exist as a sphere separate from human actions, ambitions, and needs (Johnston, 2002). Modern crisis situations (in environmental, developmental, agricultural, social or energetic aspects) are not independent of one another. It is one global crisis, which refers to the human approach to the environment, and cannot be solved within the jurisdiction of

17

individual countries. This was an elaboration on U'Thant's conclusions from the famous speech at the UN in 1969.

An attempt at summarizing human successes and failures in the 20th century, was a valuable part of the "Our Common Future" report, and became the basis for the outlined vision for the further development of mankind.

The following issues were listed as essential (WCED, 1987):

- Stabilizing the size of the human population, while emphasizing that not the population growth itself, but rather its limitations resulting from the availability of resources, are the major problem.
- Ensuring food for humans while understanding that the problem lies in its distribution, since the global amount of food is already sufficient.
- Preventing the loss of species and—more widely—genetic resources, especially regarding rainforests and areas in need of conservation.
- Energy issues, especially: energy conservation, the search for new sources, refraining from burning fossil fuels (especially in terms of global warming), renewable energy sources, controversy over nuclear power.
- Industrial issues, including resource conservation.
- Issues related to human settlements, especially in urban areas. It was anticipated (this prediction actually turned out true) that at the beginning of the 21st century, the percentage of population living in cities would, for the first time in history, exceed the percentage of rural population. This leads to a number of environmental, infrastructural and social issues: clean water, sanitary aspects, availability of healthcare, transportation, schooling, interpersonal relationships.

Among the main threats, the following were listed (WCED, 1987):

- A decrease in the areas under cultivation (main causes: soil erosion and desertification).
- Radical deforestation, especially in South America and Asia.
- Excessive burning of fossil fuels and air pollution related to it (in the global aspect, this can lead to global warming, whereas regionally it may cause acid rain, which not only threatens living organisms but also cultural monuments),
- The dependence of industry on natural resources.
- Extensive gas emissions, which threaten the ozone layer.

Also, in terms of purely human aspects, attention was drawn to (WCED, 1987):

- Increase in the number of starving.
- Increase in the number of illiterates.
- Increase in the number of people with no access to clean water.
- Increase in the number of people without healthy and safe housing.
- Increase in the number of people without firewood.
- The growing discrepancy between the rich and the poor nations.
- Arms race.

Successes were also noted (WCED, 1987):

- Decline in infant mortality.
- Extension of average lifespan.

- Increased number of people able to read and write.
- Increase in the proportion of children attending schools.
- Growth in the global food production, which exceeds the population growth.

The balance of success and failure is unclear. The discussion placed strong emphasis on the often neglected needs of the poor countries. It has been observed that, despite various aid programs, the gap between the rich and the poor countries not only was not reduced, but instead grew deeper and deeper. This is best shown by the fact that, although the number of literate people is increasing, so is the number of illiterates. Moreover, it has been emphasized that economic issues are strictly associated with environmental conditions. This is a two-way relation. Economic development has a negative impact on the environment (if only in terms of using up resources), but on the other hand, degradation of the environment may restrict economic development.

The concept of sustainable development was supposed to solve these overlapping problems. According to Brundtland's Commission "sustainable development is development that meets the needs of the present without compromising the possibilities of future generations to meet their own needs" (WCED, 1987).

Despite several similar proposals and terms, it was the report "Our Common Future" that turned out to be crucial. Its major achievement was the general acceptance of the concept of sustainable development both in scientific and in political circles, as well as in the wide circle of world's public opinion. The proposed definition (known as the principle of sustainable development) gained a normative dimension and all future development strategies referred to it.

The report "Our Common Future", although widely discussed, was not a specific strategy, which could be implemented. Such strategies were yet to be prepared during the scheduled Earth Summit in Rio de Janeiro in 1992.

Meanwhile, the world has changed, following the fall of communism in Eastern Europe. It started with the events in Poland in 1989. Meetings of the Round Table were held between the communist authorities and the opposition led by Lech Walesa and Trade Union 'Solidarity', which resulted in an agreement that enabled the introduction of a democratic system in Poland. During the deliberations, the Subdivision of the Round Table for Ecology was created. Except for energy issues (the government party supported the idea of building a nuclear power plant, unfortunately the RBMK type—see technical plane of sustainable development, the opposition was against that concept), almost complete agreement was achieved, including on the fundamental issue—the necessity to change the direction of Poland's primary development. Adoption of the principle of eco-development (this was how sustainable development was originally referred to in Poland) was postulated, as well as making changes in industrial, energy, urban, transport and agricultural policies. As the documents describe it, dying forests, water and air pollution, poisoned soils, food contamination, would all occupy an increasing proportion of the country. It is crucial to make a significant turn in describing priorities for the country's further social and economic development by adopting eco-development and eco-policy in strategic plans (Kozlowski, 2005).

The changes postulated were a turning point in this part of Europe; therefore they deserve a detailed view. They regarded introducing ecological goals into the set of social and economic goals, as well as those related to spatial planning (Zukowska, 1996). Moreover, the nationwide eco-development strategy was to be enhanced by regional and local strategies.

Among the main problem groups, the following were emphasized (Protocol ..., 2004):

- Out of the actions regarding the economy, the necessity to restructure industry (including the promotion of environment-friendly technologies), reduce the amount of waste generated, stop wasteful cutting down the forests, eliminate food contamination, reduce air pollution (including pollution from the automotive industry) and to organize water resources management.
- In the legal aspect an amendment to environmental law was proposed, which would include reference to ethical issues related to a proper shaping of the human-nature relation.
- In terms of international cooperation, the threats associated with the construction of power plants burning coal near the south-western Polish border, were pointed out (the 'black triangle'; the problem was solved in the following years), also introducing a complete prohibition on waste import to Poland was postulated.
- In the group of social issues, providing universal access to information on the environment and guaranteeing the freedom of independent ecological organizations, was requested.
- The intervention cases were also important; they regarded specific locations and industrial facilities that were assigned for immediate solution.

In the long run, the agreement reached in Poland during the deliberations of the Round Table enabled the East-European Countries to access the European Union.

Poland also participated in the scheduled UN conference in Rio in 1992.

Preparations for this Earth Summit were announced in 1989 under the UN Resolution no. 44/228. The Preparation Committee (PrepCom) began work in March 1990. A month later (28th Apr/2nd May) conferences were held in New York and in Washington D.C., which—although officially not under the UN—undoubtedly had an impact on the preparations for the Earth Summit. It regarded the issues of global protection of the environment. A group of American decision-makers, led by the author of the book "Earth in the Balance", Senator Al Gore (Gore, 1992), had invited parliamentary delegations from other countries (41 delegations arrived) to discuss the following problem groups:

- Global climate change.
- Disappearance of the ozone layer.
- Controlled development.
- Population.
- Deforestation and desertification.
- Conservation of oceans and water resources.
- Maintaining biological diversity.

It is worth stressing the work of the team for controlled development issues. They concentrated mainly on searching for ways of balancing ecological needs with economic development, therefore directly referring to one of the major problem groups regarding sustainable development.

However, most of the meetings in that period were organized directly by the United Nations. The most important conferences of the time were (Czyz, 1992):

- 2nd World Climate Conference in Geneva.
- World Conference on Industrial Environmental Management in Rotterdam.
- International Conference on Science for Environment and Development in Vienna.
- Conference on Water and the Environment in Dublin.
- Conference of the United Nations Industrial Development Organization (UNIDO) on pro-ecological industrial development in Copenhagen.
- A preparatory conference of non-governmental organizations (NGOs) in Paris.
- Regional conferences: for Europe and North America in Bergen.

It should be emphasized that, for the first time, apart from the official national delegations, numerous non-governmental organizations (NGOs)—whose role in implementing sustainable development was soon to become very important, especially at the local level—were invited to participate. Also during the sessions of the proper Earth Summit, a parallel discussion was held within the Global Forum, which gathers the representatives of over 1600 NGOs (Hannenberg, 1992).

Renewing the "World Conservation Strategy" in 1991—the document was renamed "Caring for the Earth" (IUCN, 1991)—turned out to be a significant accent. It stressed the fact that actual biodiversity protection and a proper reshaping of the human-nature relation requires prior rebalancing of the human-human relation.

The UN's Earth Summit in Rio de Janeiro in 1992 turned out to be a special event. A wide program of mankind's development was formulated there, based on the principle of sustainable development. Were it not for this conference, the concept of sustainable development would have probably become one of the many ideas, which—however interesting—remained purely theoretical constructs, with no (or little) reference to reality.

The conference in Rio took place between 3rd and 14th April 1992. This Earth Summit was organized under the slogan 'Environment and Development' and gathered representatives of 172 countries (around 30 thousand participants in total). The most evident result of the discussion led throughout the deliberations was the acceptance of five important documents (Earth Summit, 1992):

1. "Rio Declaration on Environment and Development". It is a set of principles regarding mankind's rights and obligations, a specific code of conduct towards the natural environment, to which all the other documents of the Earth Summit refer. These principles—prepared on the basis of the Stockholm Declaration of 1972 and later UN initiatives—show the primary problem groups regarding mankind's future development with more detail (and are discussed in chapter 2 of this book).
2. "Agenda 21", which concerns actions reaching the 21st century. This program of sustainable development involves the integration of widely understood economic and developmental issues, with environmental issues. The document contains numerous specific instructions for governments and international organizations, aiming at integrating global policy with the decisions made within the jurisdiction of individual countries (also at the

local level). The document has been divided into four main thematic sections (Keating, 1993):

– socio-economic aspects, i.e. the fight against poverty, changing the consumption model, demographic dynamics,
– resource conservation and management, including specific strategies, i.e. concerning the protection of biodiversity, the fight against excessive deforestation, or protection of the atmosphere,
– strengthening the role of various social groups, especially local communities and NGOs,
– ensuring the means for the realization of the intended action program, not only including financial aspects, but also the problem of slow transfer of technology harmless to the environment (this aspect aroused strong opposition from the USA), educational and scientific aspects, as well as the issues of international institutional agreements.

Moreover, under chapter 28 of the Agenda, the Commission on Sustainable Development (CSD) was founded. Its goal is to prepare and monitor the mechanisms associated with the implementation of sustainable development.

3. "Convention on Biological Diversity"—drew attention to the importance of the abundance of wildlife in ecological, genetic, as well as scientific, educational, social, cultural, recreational or esthetical dimension (Danielson, 1995).

The convention goes beyond the conservative type of environmental protection (only concerned with the most precious types of ecosystems), and beyond strict species preservation. This is because biodiversity was described as the differentiation of all living organisms possible. This concerns diversity within species, between species and between ecosystems. Moreover, the postulate of biodiversity preservation is not intended to include only natural environments, but also those processed by humans. The species was pointed as a primary determinant. Some of the species living on Earth have an impact on maintaining homeostasis of the biosphere, some are essential to humans (be it as a source of food, or in the health aspect, e.g. the issue of medicines), but the meaning of many others has not been determined so far. What's more, scientists—despite constant progress in science—are not able to even determine the approximate number of species on Earth. The estimates available range between 3 million and 30 million species, most of which have not been found yet. Simultaneously, it is estimated that around 100 species go extinct every day—and even 40,000 every year (May, 1992; Myers, 1986). The convention states that all countries have the right to use their biological resources, but they are still obliged to maintain biodiversity (especially in the case of endangered species) and ecological balance, as well as to restore the ecosystems already degraded.

The document also raises the issue of the species alien to the given environment, which could pose a threat to the ecosystem (and which are occasionally—unintentionally—transferred during passenger flights or sea cruises).

Drawing attention to the developing biotechnologies, including genetic modifications (e.g. in the food aspect known as Genetically Modified Objects—GMO's), was precursory. Creating independent supervisory authorities was postulated, in order to assess the threat associated with specific technologies.

It was also decided to create a special aid fund for the poor countries, where environmental degradation—and hence, the loss of biodiversity—often results from poverty. This is the case with cutting down tropical forests. This problem was explicated in the next document.

4. "Forest Principles—Statement of Principles for a Global Consensus on the Management, Conservation and Sustainable Development of All Types of Forests". It underlined the significance of the functions performed by the forests. The following should be listed:

• Ecological functions:
  – forests maintain ecological processes on Earth,
  – they contribute to the increase of biodiversity,
  – they are reservoirs for water and carbon,
  – they absorb carbon dioxide, which is one of the major greenhouse gases, therefore moderating adverse climatic changes.
• Economic functions, associated with obtaining wood and—in a wider range—food. It has been noted that not only natural forests, but artificial forests are valuable as well. The latter may especially serve as a source of biomass, whose combustion is one of the basic renewable energy sources.
• Social, cultural and spiritual functions, including i.e. esthetic aspects, touristic aspects as well as educational and tutorial aspects.

The declaration in question had a special meaning in Brazil, where the conference was held, and at the same time, where excessive deforestation is a major environmental problem. However, the declaration had merely the status of a recommendation, and not of a legally binding document (Hannenberg, 1992).

5. "Framework Convention on Climate Change", describing the tasks in terms of preventing global warming and its effects, especially pointing out the necessity to reduce the emissions of greenhouse gases to the atmosphere. This was an explication of the works led by other UN agendas, out of which the following should be mentioned:

• International Hydrological Programme (IHP), founded by the UNESCO in 1975 (IHP, 2009). At first it was aimed at developing technological measures for the proper management of water resources; later on, several other environmental factors were taken into account (e.g. climate changes), which eventually placed the discussion in the context of sustainable development.
• World Climate Research Centre, established in 1980 (WCRP, 2010), based on the World Meteorological Organization (WMO, established in 1950), which in turn referred to the International Meteorological Organization (IMO, whose traditions date back to 1873) and International Oceanographic Commission (IOC, founded in 1960). It is worth adding, that prior to the Rio Conference, the Center had prepared the World Climate Programme (WCP).

Moreover, the convention on climate change had identified the most vulnerable areas:
  – areas most likely to be flooded as a result of rapid melting of glaciers (lowlands, seacoasts and small islands),
  – areas subject to drought and desertification, present in all regions of the world.

It has been stated that mainly the rich countries are responsible for the excessive emissions of greenhouse gases to the atmosphere. Therefore, it should be for them to show the most concern in the recovery program. Unfortunately, this was not been exceptionally successful, due to the clear opposition from the United States.

Two of the documents mentioned ("Rio Declaration" and "Agenda 21") were passed by the UN General Assembly, the remaining conventions and declarations were signed by individual countries, although part of the delegations refused to sign some of the documents.

This was not the only problem. "Agenda 21"—the most important document of the Earth Summit—required around 600 billion dollars a year for its implementation (Baltscheffsky, 1992). This amount was never collected. As a result only some of the decisions made in Rio de Janeiro were adopted.

Despite these limitations, the importance of the conference was enormous.

First of all, the unprecedented media publicity accompanying the deliberations contributed to the wide popularization of the issue of sustainable development, due to which it ceased to be merely a domain of scientific discussions.

Secondly, through "Agenda 21", the conference had provided the model and the methodology in the preparation of strategies of sustainable development at the global, regional and local levels, therefore enhancing the recommendations of the "Our Common Future" report.

Thirdly, the documents from Rio were the subject of further work, performed by various UN organizations. The following agendas play a particularly important role here: Economic & Social Council (ECOSOC, operating since 1945), associated with the Department of Economic and Social Affair (DESA) and especially the Division for Sustainable Development (DSD), which belongs to the latter and watches over the implementation of "Agenda 21" in its various dimensions and at various levels.

Moreover, in 1990, the International Council for Local Environmental Initiatives (ICLEI) was established. After 1992—although the abbreviation remained unchanged—the name was changed to Local Governments for Sustainability, thus emphasizing the importance of the local aspect in taking action for the environment.

During the 90's, a major role was also played by the subsequent reports from the Club of Rome.

In 1993 a report was prepared titled "Beyond the Limits: Confronting Global Collapse, Envisioning a Sustainable Future" (Meadows et al., 1993). It was an attempt at summarizing all that had happened since 1972 (when the first report was published). The same authors stated that the trends outlined then were not interrupted, but instead had become aggravated. They have even suggested the occurrence of overshoot phenomenon, which is understood as an accidental, unintentional exceeding of the environment's limits. As they prove in the report, the rate of human consumption of many basic resources and of their production of various types of pollutants has already gone beyond the environment's physical capacity. However, we still have the technological and economic conditions to create a society able to survive (Meadows et al., 1993).

Another report was published in 1997. It was titled "Factor Four" (Weizsäcker et al., 1997) and proposed the 'revolution of efficiency'. The report included the question: how much do we have to increase the efficiency of our resource usage in order to maintain our present level of well-being? While searching for the answer, it had been pointed out that even now it is possible to achieve four times higher effects of resource management due to the progress in technology and, in the long run, the increase could even be ten times the present efficiency. What this means in practice, is doubling our well-being—which is very significant—at the same time reducing the consumption of natural resources by half.

The increase in efficiency proposed in the report would lead to achieving seven primary goals (Weizsäcker et al., 1997):

- A better life (concerning its quality).
- Less pollution and waste.
- Profiting from exhausting less resources.
- Using market mechanisms in accordance to—as it had been described—economic-ecological common sense.
- Multiple use of capital, due to savings (for instance, if energy-saving bulbs were popularized, there would be no need for an additional power plants, due to the reduction in energy consumption).
- International security, since even now a number of conflicts have a resource background; if consumption is reduced, the resources will be more accessible.
- Equity of work: in terms of wasting human talent in the case of high unemployment.

The study in question presented fifty specific examples of technologies that would help achieve the goal. In most cases they refer to the rich countries. For instance, at the time when the report was published, a typical American household used about 300 liters of water per day. Significant savings are possible, of course, but how do we transfer this to countries, where millions of people have no access to clean water—this makes over a billion people in the world (No water ..., 2004)? Nevertheless, concentrating on the problems of the northern countries is—paradoxically—correct, since it is those countries that use and waste the most natural resources.

The report "Limits to growth, The 30-years Update" (Meadows et al., 2004)—an updated version of the famous study "Limits to Growth" of 1972—had a wider dimension. It stated that mankind stands before an important choice between the three possible ways of development:

- Acknowledging that there are no limits. This means maintaining the present economic methods, which leads to a civilizational collapse.
- Acknowledging that, although there are limits, the people will not change their present lifestyle (especially in the case of the rich countries, also in terms of their refusing to aid the poor countries), which leads to a collapse.
- Acknowledging that there are limits, some of which have already been crossed, however it is still possible to stop the destruction of the environment, providing a radical change in economic systems, based on the idea of the common good. Realizing this scenario may prevent the collapse, although the present situation is, unfortunately, far from satisfactory.

The newest report from the Club of Rome is titled "Factor Five, Transforming the Global Economy Through 80% Improvements in Resource Productivity" (Weizsäcker et al., 2009). It is an elaboration on the report "Factor Four", showing more examples of growth in efficiency and energy saving, as well as the propositions for structural solutions, including long-term reforms of tax systems. The authors suggest that—taking into account the present level of technological development—that not a fourfold, but a fivefold efficiency improvement in industrial production is possible, without increasing the usage of resources.

Another important initiative, which goes back to the 90's, is the establishment of another "Earth Charter". Work on the document began in 1992 after the famous conference in Rio and were finished in 2000. The Charter adopted 16 principles, divided into four problem groups (Earth Charter, 2010):

- Respect for and protection of life and its biodiversity.
- Protection and restoration of the integrity of Earth's ecological systems.
- Economic and social equity.
- Democracy, preventing violence, promotion of peace and tolerance.

The preparation of the document and numerous consultations were supervised by an independent commission (World Charter Commission) and the final Charter was accepted by many important organizations, including IUCN and the UN.

Then, on 6–8 September 2000 in New York, the UN General Assembly organized the Millennium Summit, participated in by the representatives of supreme authorities from 189 countries. Two documents were passed during the deliberations:

- "The Millennium Declaration" concerning the commonly accepted values, such as freedom, equality and tolerance (Mihelcic et al., 2006).
- "The Millennium Development Goals Report" concerning the need to end poverty and hunger, universal education, gender equality, child health, maternal health, combat with HIV/AIDS, environmental sustainability and global partnership (Annan, 2000). This report is updated annually.

This was similar to the Earth Summit in Rio, where—among the prepared documents—a general declaration regarding ethical values ("Rio Declaration") as well as a specific action program ("Agenda 21") were accepted.

The significance of the Millennium Summit is shown by the fact, that it was preceded by the publication of a study, signed directly by the UN Secretary-General Kofi A. Annan, titled "We the Peoples: The Role of the United Nations in the 21st Century" (Annan, 2000), later discussed within the General Assembly. It is worth emphasizing, that for the first time, the discussion was so clearly put into the context of challenges, brought about by the ongoing process of globalization.

Among its advantages, the following were distinguished:

- The facilitated removal of barriers in trade and in the cash flow.
- Support for technological progress.
- Boosted economic growth.
- Improving living standards.

Inequity has been recognized as the biggest problem. After all, the advantages of globalization are limited to the small group of the rich countries, whereas its costs are borne by everyone. Annan's report suggests the 'inclusive globalization' (more details on globalization can be found further in this book) as a future proposal, which includes a fairer distribution of wealth. Also, the tasks (very close to the Millennium Development Goals) for the succeeding years were listed (Annan, 2000):

- Reducing poverty and famine by half by the year 2015 (this, however, will probably not be possible). One of the supporting proposals is the UN project 'Cities Without Slums'. It was also stated that every action for the reduction of poverty is also a step toward preventing further military conflicts.
- Increasing the number of people with higher education.
- Providing employment, especially for young people. Already in 2000, 80 million people could not find work, 80% of which was in the developing countries as well as those in the so-called interlude.
- Supporting gender equality (issues about the availability of a labor market were stressed).
- Improving health status (especially in terms of reducing child mortality).
- Improving the health of parturient women (reducing perinatal mortality).
- Supporting democracy (including the fight against corruption).
- Availability of fundamental elements of infrastructure, necessary in households (e.g. the issue of clean drinking water).
- Ensuring sustainable development, especially in terms of national sub-strategies.

Each of the countries was obliged to prepare its own sub-strategy, which would include the local conditions. In practice, the same problem was encountered as in the case of the already mentioned "Agenda 21" of 1992. Collecting the funds necessary for the realization of the planned actions (about 50 billion USD), turned out to exceed the capabilities of the UN. Therefore, preparing annual reports was necessary, in order to show up-to-date what had been achieved, and to what extent.

Similar challenges were presented in the report by Federico Mayor, ex-Director-General of the UNESCO, titled "Future of the World" of 2001 (Mayor & Binde, 2001).

Another Earth Summit—which took place in Johannesburg on 26 August—4 September 2002 (Earth Summit, 2002), almost precisely 10 years after the famous conference in Rio—referred to these studies. This time as well, the interest in this meeting was enormous, representatives from 190 countries (about 50 thousand participants in total) being present.

During this Summit, two documents were prepared, signed by heads of states (Al-Hadid, 2002):

- "Johannesburg Declaration", which directly concerned the "Rio Declaration" and called for implementing sustainable development.
- "Action Plan", enhancing the discussion over "Agenda 21" by the "Report on Millennium Development Goals", passed during the Millennium Summit in 2000.

Was the conference in Johannesburg a success? Certainly, the climate accompanying it was different from that of the Earth Summit in Rio.

First of all, it took place only a year after the terrorist attack on New York. This changed the way of thinking of many world leaders, pushing the environmental issues to the background.

Secondly, in Rio, the creation of large scale strategies was supported. These encountered serious problems during their realization, mainly related to lack of sufficient funds. In Johannesburg, solving individual problems, easier to fund, was opted for. The American proposal was interesting, as it included action within the following fields: water for the poor, clean energy, fight against poverty and famine in Africa, forest preservation, common fight against AIDS, tuberculosis and malaria (Clarke, 2002).

Thirdly, the deliberations occasionally took on the nature of an ambitious confrontation, especially between the United States and the European Union. Achieving a consensus in such a situation is extremely difficult.

Moreover, under the pretext of the necessity to reduce the global population, the representatives of the EU, Canada and some feministic organizations promoted the inclusion of not only all forms of contraception but also free abortion, into the packet of basic health services. This proposal was protested against by the Vatican, the USA, Ireland, Spain, Italy and by the Third World countries. Eventually, the final version of the document only stated that basic health services are subject to local regulations, in force within individual countries, with respect both to cultural and religious traditions (Clarke, 2002).

The final evaluation of the Summit is not easy. The proposal of implementing specific solutions in place of huge strategies seems more realistic. Yet none of the previous Earth Summits was able to stop the degradation of both the natural and social environment.

Independent annual reports on the state of the world, published since 1975 and prepared by the Worldwatch Institute, may be a significant tip for the future. According to the principle of sustainable development, the latest of them not only include environmental issues but also economic and social issues. The studies from recent years are summarized below (Worldwatch, 2010):

- "State of the World 2000", concerning the criticism of global economy, threatening the planet's environment. The authors stress that, although in the 20th century, humans learned how to travel to the Moon, create sophisticated computers and modify human genes, the major challenges still include: providing clean water; preventing loss of biodiversity; and reducing the emission of pollutants (especially those related to the development of coal power), which cause climate changes.
- "State of the World 2001", presenting the vision of a sustainable economy, whose implementation might ward off the destruction of Earth. Particular emphasis was put on environmental degradation associated with the fast-paced development of the rich North and the growing impoverishment of the rest of the world.
- "State of the World 2002", raising global issues, associated with the Earth Summit in Johannesburg. The work contains a significant statement that, although 10 years have passed since the Earth Summit in Rio, we are still far from achieving even the basic goals adopted then.
- "State of the World 2003", on the need for (civil, social, but also governmental or even corporative) change, in the face of advancing biodiversity loss, threats associated with the global warming, or the still unsolved problem of poverty, experienced by millions of people in various parts of the world. The fields in which changes have already been made, were also pointed out (e.g. reducing population growth in many countries, or the promotion of alternative 'green' energy sources).

- "State of the World 2004, Special Focus: The Consumer", devoted to increasing consumerism and the negative consequences of maintaining this trend.
- "State of the World 2005: Redefining Global Security", raising the issues of global security. Apart from the discussion on the level of environmental degradation, or potential threats associated with the spread of diseases (particularly in the poor countries), important issues of problems resulting from the competition between the rich countries concerning access to oil and other crucial resources, as well as the conditions of terrorist activity, were also discussed.
- "State of the World 2006, Special Focus: China and India", which stressed the global consequences of India's and—in particular—China's rapid development, combined with the fast-paced increase in their demand for resources.
- "State of the World 2007: Our Urban Future, regarding cities". The subject matter was the consequences of an increasing urbanization process, both in the rich and in the poor countries.
- "State of the World 2008: Innovations for a Sustainable Economy", which states that further human development depends on transforming the economy on the basis of sustainable development. An economic dimension of implementing particular solutions is also shown, e.g. the innovations introduced at DuPont led to a radical reduction in greenhouse gas emissions by the company's facilities, which brought about savings of as much as 3 billion dollars.
- "State of the World 2009: Into a Warming World", on the consequences of the global warming and the opportunities to limit them.
- "State of the World 2010: Transforming Cultures", pointing that preventing a global ecological catastrophe is only possible through making a turn in the direction of sustainable development.

Apart from the above studies, regarding the conditions of development at the global level, undoubtedly much depends on the most important regional 'actors' such as the European Union; and it is the European aspect of the debate on protection of the environment and sustainable development, which will now be discussed.

2 EUROPEAN PERSPECTIVE

Early historic European initiatives for protection of the environment have been presented during the discussion on the global perspective of the problem.

The current aspect refers to the cooperation—started in the mid 20th century—which resulted in the creation of the European Economic Community (EEC)—see Table 4. It is true that these actions were mainly aimed at the anticipated economic benefits not on natural protection of the environment (with time, however, the word 'economic' was removed and EEC become EC). Despite such conditions, the legislation adopted is impressive (Klemmensen et al., 2007; Baker, 2000).

Passing of "The Single European Act" in 1987 is considered a turning point. However, the modern cooperation for the protection of European environment began earlier—in 1972, during the already described Stockholm Conference and with the establishment of the Helsinki Commission (Anderson, 1997).

Table 4.   Important European documents regarding protection of the environment and sustainable development (Author's own work).

Year	Name of the document
1957	"Treaty of Rome" on Establishing the European Economic Community.
1987	"Single European Act".
Since 1973	"Environment Action Programmes" (6 editions so far).
1993	"Maastricht Treaty" (also known as "Treaty on European Union").
1998	"Treaty of Amsterdam".
2000	"Lisbon Strategy".
2001	"EU Sustainable Development Strategy".
2006	"Renewed EU Strategy for Sustainable Development".
2007	"Treaty of Lisbon".

These were UN initiatives, but in the very same year, the European Community introduced this problem into the internal discussion during the Paris Summit. It was the meeting of the heads of states, which clearly stressed that economic development must include the principles of protection of the environment.

During the deliberations, the "Environmental Action Programme" was also prepared. This initiative later turned out to be one of primary European instruments in environmental protection policy. Subsequent programs were related to each other, hence introducing another program is impossible without taking into account the conclusions from the previous one (Klemmensen et al., 2007).

The first "Environment Action Programme" covered the years 1973–76 and regarded the need for improving the state of natural environment as well as improving the health status of Europe's residents.

The main postulates adopted were:

- To reduce the emissions of pollutants.
- To maintain the ecological balance and ensure the safety of the biosphere.
- To manage natural resources rationally.
- To take action for the improvement of the quality of life.
- To support regional and international initiatives for protection of the environment.

The statement that economic expansion must not be the goal in itself, but should also be expressed in the improvements in the quality of life, was very important. This is one of the conditions, taken into consideration today, as a part of the discussion on sustainable development.

Another "Programme" covered the years 1977–1981 and updated the list of specific actions, necessary for the basic environmental protection.

The "Third Programme" was introduced for the years 1982–1986. It concentrated on supporting the actions preventing pollution and environmental degradation. It was, therefore, not only the recovery from the effects of environmental problems, but also an attempt at doing away with their causes. Moreover, an integration of protection of the environment policy with other sector policies was attempted.

The "Fourth Programme" regarded the years 1987–1992 and continued the support for environmental pollution preventing strategies.

The basis for the "Fifth Programme" was the discussion held during the Earth Summit in Rio de Janeiro 1992. The program covered the years 1993–2000. Even its title is characteristic: "Towards Sustainability". The report recalled the definition of sustainable development, pointing out the obligation to sustain the environment for future generations. This task was to be realized through mutual integration of social, economic and ecological issues, especially in relation to decision-making processes. The significance of joint responsibility had been emphasized as well, regarding both central and local authorities, as well as social organizations.

The three main goals included (Klemmensen et al., 2007):

- Maintaining the general standard of living.
- Maintaining access to natural resources.
- Avoiding permanent damage to the environment.

Sector goals had also been pointed out, including industry, energy, transport, agriculture and tourism. These were linked with, for example, the issues of climate change, the quality of air, the development of urban areas, water resources management and biodiversity preservation. This program largely referred to the instructions of "Agenda 21", a document which was a model strategy for implementing sustainable development. Although the original Agenda within the UN was only partly funded, the Life program prepared within the EU was more successful. The contributions within this program were calculated depending on the size of a particular country and its national income and, since 1996, also countries associated with the EU are included in the program (Borys, 1999).

Then the Sixth Environment Programme "Our Future, Our Choice" has been introduced, covering the period until 2012 (Andersson, 2003). It confirmed the significance of the goals realized so far, but also unified other documents and legal acts. The most important part of the program was to determine the seven fundamental strategies (Our Future, 2001). The first four strategies were already adopted in 2005; the rest was introduced in 2006.

- "Strategy on Air Pollution" with a time span prolonged to 2020. It emphasized the significant reduction of pollutant emissions, which, however, is still recognized as insufficient. Therefore, the main goal is to further reduce the level of emissions of substances dangerous to human health, including fine particulate matter and tropospheric ozone. According to the authors of the strategy, the presence of this type of matter (PM2,5) in the air shortens the life expectancy of Europeans by about 8 months, resulting in as much as 3.6 million deaths per year.

  The need to reduce natural areas—especially forests—threatened with degradation as a result of excessive pollution of the atmosphere (i.e. the issue of sulfur dioxide and nitrogen oxides, which cause acid rain) is also stressed.

- "Strategy on the Prevention and Recycling of Waste". This was the European Commission's reaction to the increasing waste production. This problem especially refers to municipal waste, 49% of which is not recycled in any way and goes straight to landfills. In order to counteract this, a waste hierarchy was adopted. It puts the main stress on preventing waste generation (e.g. in terms of possible technological changes) and on the promotion of recycling, which enables processing the used materials (mainly in terms of basic waste streams of municipal waste, associated with packages, used electronic hardware and unused vehicles).

- "Marine Environment Protection Strategy", which not only includes improving water quality, but also the preservation of fish stock and other marine organisms. Part of the program was setting up local European Marine Regions.
- "Strategy on the Sustainable Use of Natural Resources", including fossil fuels, metals, minerals, wood, as well as plants and animals. These resources are crucial for ensuring further economic and social development, but are also consumed too quickly (this is a direct reference to the principle of sustainable development, pointing at our obligation to secure the resources for the future generations). Moreover, in case of fossil fuels, their combustion is a threat to the whole natural environment (regional pollution and global warming). Therefore, it is necessary to use the resources more effectively (especially in terms of energy production and transport), hence reducing the level of exploitation.
- "Strategy for Soil Protection", which emphasizes the various functions of soil. These include i.e. supplying humans with food, biomass, resources, or pollution filtering. Moreover, it is a key component of the environment, which contributes to natural cycles of water, coal and nitrogen. This resource is practically non-renewable due to the long time of generation of the soil layer. It is also a biodegradable resource. This can include physical aspects (wind and water erosion), chemical aspects (acidification, salinization, local pollution) and biological aspects (reducing the level of organic matter, viral and bacterial contamination). The scale of soil degradation brought about the necessity to prevent this negative phenomenon. Already 12% of Europe's land area (115 million ha) is threatened with water erosion, and 6% (42 million ha)—with wind erosion. Moreover, up to 45% of soil is characterized by a low content of organic matter. In the face of such threats, our way of using soil must be changed and an intervention at the source is needed, also in other than agricultural aspects of human activity, which contribute to soil pollution.
- "Strategy for Sustainable Use of Pesticides" states that, although very helpful in fighting vermin, pesticides may be dangerous to humans. Therefore, their use should be limited, and other solutions should be sought (e.g. using less dangerous substances or introducing crop rotation).
- "Strategy on Development of Urban Areas", within which ecological, economic and social aspects of sustainable development are underlined. Apart from discussing the indicators of a clean environment, the issues of a proper lifestyle, transport, infrastructure, construction, noise and spatial planning were also analyzed.

The strategies described above—as already been pointed out—are part of the Sixth Environmental Action Programme. So far, it is the latest program introduced by the European Union. An analysis shows that it mainly refers to the part of sustainable development issues, which is related to traditional environmental protection. This does not mean, however, that other aspects of sustainability are neglected within the EU. The recommendations of the Fifth Environment Action Programme, which was targeted at this type of development, are still in force. Moreover, the Community's initiatives are not limited to the program mentioned above.

Also the "Single European Act" (Kiss & Shelton, 1993), introduced in 1987 (prepared a year earlier)—which was an enhancement to the "Treaty of Rome" of 1957—was a crucial moment. The Act was mainly devoted to creating a common market and removing internal trade barriers in the European Community, it also included reference to protection of the environment. It postulated i.e. the need to support scientific and technological basis for the protection of human health, as well as to ensure a reasonable and rational use of natural resources (Single European Act, 1987). Also, the obligation has been pointed out, to make an economic balance of profits and losses, which may result from taking—or refraining from taking—action in the environment.

Three years later, when the principle of sustainable development, had already been introduced to the UN documents, during a conference of the countries belonging to the EEC in Bergen, the will to implement sustainable development strategy in all member states has been expressed.

Important treaties followed this declaration (Agenda 21, The First ..., 1997):

- "Maastricht Treaty", which entered into force in 1993 and led to the establishment of the European Union (hence the other name—"The Treaty on European Union"). It imposed introducing requirements on protection of the environment to all implemented policies in order to realize sustainable development.
- "Treaty of Amsterdam" of 1998, which set sustainable development as one of Europe's major goals.

In view of these documents, the EU priorities are as follows (Agenda 21, The First ..., 1997):

- Internal integration with less developed regions.
- Integration between the EU and its neighbors.
- Integration of environmental policy in reference to other developed countries.
- Integration of environmental policy in reference to developing regions.

After passing "The Maastricht Treaty", in 1994 the European Environmental Agency (EEA, 2010) was established. Its main priority is to perform scientific research, gather information, releasing it to the public opinion and to the institutions responsible for protection of the environment. Preparing regular reports, which make an overview of the state of the environment in Europe, is an important way to realize this task. The first of such reports was titled "Dobris Assessment" and was released in 1995 (Andersson, 1997).

Apart from the initiatives listed above, the recommendations (not legally binding, though), regulations, decisions and environmental directives also have an impact on the realization of protection of the environment goals and implementation of sustainable development. The directives, which make the framework for an integrated approach to the environment, are the most important—see Table 5.

Also the energy and climate package, in use since 2007, plays an important role. Its general motto is '3 times 20 before 2020', which means that before 2020 there should be (Globe-Europe, 2010):

- a 20% increase in energy efficiency,
- the share of energy from renewable sources should reach 20%,
- and the emission of greenhouse gases should be reduced by at least 20%.

Table 5.   Integrated Environmental EU Directives (EC, 2010).

Directive	Issues
No. 62 of 1996	Framework Directive on the Assessment and Control of Air Quality.
No. 60 of 2000	Framework Water Directive.
No. 1 of 2008	Directive concerning integrated pollution prevention and control.
No. 98 of 2008	Waste Framework Directive.

Environment management systems, such as ISO or the EU EMAS, refer to these documents. These are the assessments of a particular unit's (industrial facility, a specific product, but also a city) impact on the environment, including the indicators referring to the proposed improvements (these solutions are discussed further in this book).

Moreover, the "Lisbon Strategy", passed by the European Council during the deliberations in Lisbon in March 2000, is a key document (EU, 2000). The program aims at dynamizing Europe's economic development. The strategy, therefore, raises the issues of actions and measures of increasing competitiveness on the market and, at the same time, stresses the importance of creating new job possibilities. The creation of an information society was recognized as a factor supporting the realization of these goals, as well as technological innovation, associated with a dynamized scientific research, along with their practical applications. However, the strategy did not bring the anticipated results. The expected acceleration, which would enable reaching the US level of development, will surely not be achieved so quickly.

The "Lisbon Strategy" was enhanced in the aspect of environmental issues, in Göteborg in 2001, by the document: "A Sustainable Europe for a Better World: A European Union Strategy for Sustainable Development". This was also a form of the community's preparation for the already discussed Earth Summit in Johannesburg. The new version of this strategy was released in April 2006. If the 2001 document was evaluated as the 'pillar of ecology', added to the "Lisbon Strategy", the one of 2006 was formally declared as a primary strategy (Michnowski, 2007). This changed the meaning of the "Lisbon Strategy": the goal to 'catch up' with the USA was replaced by the imperative to implement sustainable development.

These documents confirmed the legitimacy of implementing many European legal solutions for sustainable development. It has also been pointed out, that the actions leading to achieving the goals of one sector policy had far too often undermined the progress in another sector. Therefore, the strategy was to introduce a hierarchy of goals (the 'cohesion policy'). This seems especially significant in the situation, when tendencies contradictory to sustainable development still prevail, the models of production and consumption are inappropriate, and the approach to politics is often disintegrated (Sztumski, 2009). Therefore, it has been stressed, that economic, social and environmental goals stimulate one another and should be realized jointly.

Among the main goals of the strategy, the following were pointed out (Kozlowski, 2005):

• Protecting natural environment and preserving its potential.
• Promoting the principles of democracy, with respect to cultural diversity.

- Supporting economy, which would use the natural resources rationally.
- Supporting sustainable development—not only in the EU internal policy, but also in foreign affairs.

From which, out of the most important long-term challenges, the following have been listed (Kozlowski, 2005):

- The necessity to limit climate change, reduce the level of greenhouse gas emissions, and increasing the range of 'green' energy use.
- Striving for the improvement of public health.
- Actions leading to improved resource management (sustainable production, consumption, and the issue of waste).
- Improved functioning of the adopted transport system and spatial planning.

The goals set for the strategy were to be realized within ten thematic areas, which include: economic development, poverty and marginalization, the ageing of society, the health of the society, climate changes and energy, models of production and consumption, natural resources management, transport, joint responsibility (co-management) and global partnership.

The presented goals and challenges clearly show that the problems are raised in a wide, interdisciplinary context. In the face of practical obstacles in realization of the level of economic development included in the "Lisbon Strategy", however, it is not clear, whether all these aspects will still be so strongly stressed in the future programs. It depends e.g. on advisory boards, among which the European Environment and Sustainable Development Advisory Council (EEAC) plays a major role. Interestingly, originally the name contained no reference to sustainable development. When it was introduced, the old abbreviation EEAC was left unchanged. The Agency prepares annual reports on implementing sustainable development in the EU, as well as consultation documents regarding sustainable development—not only throughout the whole EU, but also, in the case of actions of individual governments, for regional and local authorities.

There is also the issue of the primary legal act—the Euroconstitution, which should include references to sustainable development. On 29 October 2004, the leaders signed the treaty establishing a Constitution for Europe, however it was not ratified by all member states (the omission of Christianity's contribution to the European cultural heritage in the document's project seems to cause the most controversy) and finally replaced by already mentioned the "Treaty of Lisbon" (Unizar.es, 2011).

However, even without a common constitution, the EU legislation regarding protection of the environment and sustainable development should be highly rated.

Paradoxically, despite actions taken throughout the recent decades, the gradual, yet serious degradation of the general state of the environment within the Community continues. Pollution levels have been reduced, the quality of life has been improved, yet over 33% of vertebrates and over 22% of plant life within the EU are seriously threatened (Fines, 1999). Reversing this trend is a major challenge for the future.

The selection of documents, legal acts and important initiatives, discussed in this part of the book, shows their strong inveteracy to solutions already known to history. For instance—the Galician acts, prepared by Polish members of Parliament, were very modern in the years 1868–1874 and still make a huge impression. What is also significant, their regulations were commonly respected, which cannot be said of numerous modern initiatives, even those better formulated.

The problem lies in concentrating on environmental conditions while, at the same time, omitting other aspects of sustainable development. This has a historical background: protection of the environment has been widely discussed since the mid 20th century, whereas sustainable development has only been promoted since late 1980's. The concept of sustainable development, however, is much more inter-dimensional than protection of the environment.

Fortunately, many legal acts are being passed that take into account the wide, interdisciplinary character of the concept of sustainable development.

Regardless of the substantive evaluation of the existing legal acts, it should be noted, that—so far—in practice, they were not able to stop the degradation of the environment at the global level. This task has only been realized at the local level, mainly in areas where the level of unspoilt nature is still quite high.

Therefore, the success of sustainable development depends directly on whether we are able to achieve more and as soon as possible.

CHAPTER 2

Theoretical basis for sustainable development

Theoretical reflection on the issue of sustainable development starts with an attempt at coping with its definition. The principle of sustainable development, known from the "Our Common Future" report, is not the only attempt at explaining the notion.

1 THE NOTION OF SUSTAINABLE DEVELOPMENT

The category of 'development' is very popular nowadays, although its range of content may need to be discussed. It plays an enormous part in the economic sciences, especially in the context of economic growth. It is also significant within other problem groups regarding sustainable development, e.g. ecological (in terms of genetics, for instance: the species not only become extinct, but also give rise to new species; moreover, several stages leading to the state of ecological balance are distinguished within the ecological succession) and social, e.g. the development of societies (i.e. the distinction between those more and those less well developed), as well as changes in the individual personality of a human being and their relations with other people.

Moreover, a number of general features are assigned to the category of development (Piontek, 2005): intensiveness, dynamism, rapidity, pace; on the other hand there are: extensiveness, slowness, or duration and sustainability. It can also refer to general civilizational changes as well as more detailed issues such as science, culture, language, economy and society.

Generally speaking, it can be said that 'development' is connected with 'progress',[2] and so is a change of state of a given structure (in a civilizational

---

[2] The mutual relations between 'development' and 'progress' may be discussed. Some studies take development as undirected changes, whereas progress is a change that is directed and axiologically positive. In the era of sustainable development the two notions have become extremely close to each other and some studies treat them interchangeably.

It is worth mentioning that E. Fromm had suggested adopting the concept of 'the religion of progress', which would be characterized by: unrestricted production, absolute freedom and unlimited happiness (Fromm, 1976). It is also noteworthy that Fr. Jozef Bochenski (1994) claimed 'the faith in progress' to be one of the most harmful superstition. According to this author, progress, although it occurs in the world of nature (which results from the theory of evolution) does not take place in the world of humans (when we treat the world as a wholeness).

sense, it would be the whole of a society's activity: be it aware of this or not) that is thought of as desirable (better, closer to perfection) in the given conditions, based on a set of criteria (Borys, 2005). Consequently, 'regress' is an undesirable (worse, less perfect) change of state of a given structure, based on the same criteria. At the same time, they need to have a normative character with a very specific axiological aspect (Borys, 2005). It should include both materialistic as well as spiritual aspects. In both cases it can be assumed that a change for the better is expected. In terms of materialistic values, usually a more complicated state (e.g. improved machines) will be recognized as progressive. In terms of the spiritual aspect, it need not be such—a turn to simplicity may be presented as more desirable (a commonly known slogan: through simplicity toward perfection, the value of asceticism). Moreover, the interactions between the two fields are of significance, i.e. 'to have' or 'to be', or maybe 'to have and to be'?

The above statements indicate that the category of 'development' is associated with numerous problem groups, out of which philosophical issues are of great importance (see chapter 3). And what about sustainable development?

Some studies suggest describing this notion as primary and indefinable. In my opinion, such an approach is incorrect, since the concept of sustainable development is already too ambiguous—hence the inclusion of an ordering definitive aspect has become necessary.

This new concept has first been presented in English literature (Seuring, 2008; Newman, 2007). Further discussion in various national languages, was based on translations of original documents. It is worthwhile to present a couple of versions of the notion, discussed in Poland (Zablocki, 2005), where—apart from 'sustainable development'—the discussion also takes place on 'durable development', 'self-supporting development' or 'eco-development'. Some studies treat these notions as equivalent and interchangeable; others make a clear distinction and assign them to narrower problem groups. In fact, the type of development in question is sustainable, self-supporting and durable, all at the same time (although not every type of durable development is also sustainable).

Sustainability is expressed i.e. in the structural aspect of a given system and means reaching a state of equilibrium between its components, e.g. the actions taken within separate fields of sustainable development must not lead to degradation of the bio-social system.

In the case of durability, its main characteristic is measured in time. If a given system had been functioning in the past, is still functioning now and nothing indicates it could be damaged—that means it is durable in time. Time is also an important factor when it comes to the devastation caused by humans to the environment. In some cases, it is visible almost immediately, but often—especially when it comes to health issues—it becomes observable after a long period of the 'hibernation' (Pawłowski, 2008). Moreover, the aspect of periodicity is also important, e.g. the cycle of duration of a given organism or phenomenon (Poskrobko, 2005).

Self-support of development, in turn, is associated with the dynamism of life. This includes securing the reserves (e.g. energetic reserves), that not only would support the present-day status, but also allow taking up new challenges as well as fostering creativity, which creates stimulation to further development.

'Eco-development' is a narrower term. Clearly, the prefix eco- only points at associations with purely ecological aspects of the problems discussed. Also, the word 'ecology' is currently used in so many meanings—even an ecological way of life is mentioned (Pawłowski, 2000)—that referring it to sustainable development as well seems admissible.

It is also worth mentioning other terms, used in some publications, such as 'sustainable existence' (Blazejowski, 2007), or 'eternal duration'. Although the first term is not controversial, the second one is incorrect, since nothing lasts forever; even the duration of the Universe is not infinite (Constanza & Patten, 1995).

The most commonly known definition of sustainable development was accepted by the UN in the report "Our Common Future" of 1987. As It was already mentioned, It combined this type of development and the rights and needs of the present and future generations. Numerous other definitions, available in the scientific literature, have appeared. Table 6 lists a selection of 50 definitions of the development in question. Although almost all quoted authors accept the general UN definition, their personal insight is still worthy of attention. The majority of the given definitions stay around ecological, economic and partly social issues. Moreover, political issues are also accentuated, as well as scientific ones in some cases. Among them most interesting are proposals, which include the wider context:

- describing sustainable development as an integration of orders (T. Borys),
- indicating self-improvement of the individual, instead of civilizational development, as the goal of sustainable development (M.K. Byrski, R. Constanza),
- postulate of redefining the relations between man and nature (C. Cuello Nieto, A. Papuzinski),
- including ethical conditions in the definition, not only in the context of intergenerational equity (e.g. C. Cuello Nieto, K Dubel, D. Szaniawska, A.R. Szaniawski), but also as an ethical system within the group of systems associated with this development (E. Barbier) as well as understanding sustainable development as not only a social, but a philosophical idea as well (B. Poskrobko).

Table 6. Selected definitions of sustainable development (Piontek, 2002; Ryden et al., 2003; Sörlin, 1997; Chyrowicz, 2006; Rassafi et al., 2006; Cello Nieto & Durbin, 2008; Poskrobko, 1998; Ney, 1994; Hull, 1993; Papuzinski, 1998; Michnowski, 1995; Borys, 2005; Minisch, 1995; Szaniawska & Szaniawski, 1995; Seidel, 2006; Berdo, 2006; Livermann et al., 1998).

No.	Author (year)	Definition of sustainable development
1.	T. Bajerowski (1998)	A process (not a state), whose steering is based on minimizing losses and maximizing benefits (optimization process).
2.	E. Barbier (1987)	Sustainable development must include the goals of three systems: – biological system (biodiversity, biological production), – economic system (satisfying fundamental human needs), – cultural-ethical system (social equity, preserving cultural diversity).

(Continued)

39

Table 6. (*Continued*).

3.	J. Berdo (2006)	Providing a good life, while preserving biodiversity, social equity and sufficient natural resources. A good life does not stand for material wealth or luxurious living conditions, but for the fact that people are happy.
4.	W. Bojarski (1990)	Constant socio-economic development with respect to and with the use of natural resources.
5.	T. Borys (2005)	Sustainable development as an integrated order. An integrated order is a consistent, simultaneous creation of social, economic and environmental orders (depending on the level of management, social order may give rise to institutional-political order, whereas environmental order may give rise to spatial order).
6.	W. Burchard-Dziubinska (1994)	A development of (social, economic and natural) systems, which guarantees their harmony in such a way as to fully preserve the biodiversity.
7.	M.K. Byrski (1996)	A model of durability, which would ensure a decent existence to all residents of the globe and prevent self-destruction of human civilization. The goal of such development is self-development of the individual and not civilizational development in itself.
8.	H.C. von Carlowitz (1713)	Probably the earliest use of the term sustainable development in history. The author referred to forestry. This was intended to be a method of forest management, in which only such a number of trees is cut down, as will be replaced by growing ones, in order to preserve the forest as a whole.
9.	M.E. Colby (1990)	A strategy, which derives from natural processes of a particular bioregion, leaving vast areas of intact, virgin locations.
10.	R. Constanza (1992)	Sustainability is a relation between dynamic human economic systems and dynamic—although slower—ecological systems, within which: – human life may develop, – humans may develop, – human culture may develop, – effects of human actions are under control and cause no disturbance to the diversity, complexity and functionality of the ecological life-supporting system.
11.	C. Cuello Nieto	Sustainable development requires meeting the following conditions: – cooperation between everyone concerned at the local, regional and national level, – redefining the relation between humans and nature, – intergenerational equity, – redistribution of goods and opportunities at the global level, – not exceeding nature's capacity to regenerate,

(*Continued*)

Table 6. (*Continued*).

	– self-dependence of particular societies, – moreover, a dialectic conjunction of theory and practice must take place.
12. M. Dönhoff (1992)	Maintaining the wealth by introducing economic and non-aggressive technologies to the development.
13. M. Dönhoff (1992)	Further economic development, which includes taking into account natural resources and the necessity to reduce the harmfulness of human activity, since it threatens prosperity.
14. K. Dubel (1998)	Such a way of natural resources exploitation, investment realization, creation of technology, as would increase economic, natural and social bases of satisfying the needs of present and future generations.
15. A. Hopfer (1992)	Performing any kind of economic activity in harmony with nature, in order not to cause irreversible damage to the environment, or such management that is ecologically permissible, socially desirable and economically justifiable.
16. Z. Hull (1993)	A process, in which the reshaping of the world would not collide with decent conditions of existence and development of non-human life forms, and which would concentrate on developing and satisfying higher (spiritual) human needs.
17. S. Kozlowski (1997)	A development, which recognizes the priority of ecological requirements, which must not be disturbed by civilizational growth or economic development.
18. S. Kozlowski (1997)	A development of decent existence, including the element of self-limiting of individuals and of whole societies.
19. S. Kozlowski (1997)	All activity, which causes no degradation to the natural environment while improving human living conditions.
20. R. Kreibich (1995)	The necessity to reshape those fundamental parameters of development, which would guarantee a lasting existence to humans.
21. D. Livermann, M.E. Hanson, B.J. Brown, R.W. Merideth Jr. (1988)	Mankind's unlimited existence (with a quality of life above mere biological survival) by maintaining (conservation) fundamental life-supporting systems (air, water, soil, biosphere), the existence of infrastructure and institutions, which distribute and preserve the components of this system.
22. S. Lojewski (1998)	A lasting development of particular spatial systems, economically, socially, ecologically and spatially balanced, based on creating more and more complex and efficient, multifunctional spatial systems, characterized by high economic, social and ecological efficiency.
23. L. Michnowski (1995)	Sustainable development is complementing and strengthening the natural process of evolution by its intellectualization. It also means maintaining global society in a relative dynamical balance.

(*Continued*)

Table 6. (*Continued*).

24. D. Miethl (2002)	Sustainable development as realization of the imperative: act in such a way as to ensure that problems caused by the method of solving initial problems would not outgrow the original problems to be solved.
25. J. Misch (1995)	Sustainable development regards renewable resources, ecosystems' absorbing capacity and ecological risk, maintaining healthy biological systems and preserving the diversity of species, as well as maintaining a cultural landscape.
26. R. Ney (1994)	Sustainable development is a compromise between our civilization's pursuit for developing with minimal cost and the interest of the widely understood environment, which is being degraded by previous civilizational development.
27. D. Orr (1992)	Two aspects of sustainable development: – technical sustainability (technology helps protect the environment), – ecological sustainability (postulates changes in human behavior, reducing the level of consumption).
28. R. Paczuski (2002)	A direction in economic policy of a state based on rational management of natural resources and disallowing disruptions of balance between economic development and the state of the natural environment.
29. R. Pajda (1998)	Such a direction of economic development and social development related to it, which enables the state of the environment to be maintained, or even its restitution. It also means lack—or a significant reduction—of negative irreversible phenomena, taking place in it, especially regarding long-range effects.
30. A. Papuzinski (1998)	A new concept of an economic and social order, which aims at preventing excessive exploitation of natural resources, initiating changes counteracting the present tendencies that are harmful to the environment and seeks to change previous social patterns in human-nature relations.
31. D. Pearce, E. Barbier, A. Markandya (1990)	Realization of a specific group of desirable goals, namely: increasing real income *per capita*, improving health status and food availability, equal access to natural resources and improving the level of education.
32. F. Piontek (1997)	Ensuring a lasting improvement of the quality of life of present and future generations, which refers to permanent and integral connection between ecology and the quality of life and human environment issues.
33. F. Piontek (1997)	Permanent improvement of the quality of life of present and future generations by shaping proper proportions in managing three types of capital: economic, human and natural capital.
34. Poland: The Constitution (1997)	The Republic of Poland guards national heritage and ensures environmental protection according to the principle of sustainable development. This principle has been defined in the "Act on Environmental Protection".

(*Continued*)

Table 6. *(Continued)*.

"Act on Environmental Protection" (2002)	Sustainable socio-economic development is one, in which there is a process of integration between political, economic and social activity with preserving natural balance and fundamental processes in nature, in order to satisfy the basic needs of individual communities or individuals in present and future generations alike.
35. Polish Ecological Club (1987)	Sustainable development as realization of the following postulates: – economic development within the limits of the capacity of ecosystems, – concern for the environment should be included in the performed economic activity, – using natural resources in such a way, as to make them available to the future generations, – national self-sufficiency in terms of food, energy, water and other natural resources, – striving for a balanced, harmonic development in all regions of the country, – social and economic development based on structures mainly operating in small scale.
36. Polish Institute For Ecodevelopment (1990)	Social and economic changes occurring with: – maintaining the capacity to regenerate renewable resources, – efficient use of non-renewable resources and willingness to replace them with renewable substitutes, – gradual elimination of hazardous and toxic substances from economic processes as well as other applications, – reducing the burden for the environment and not exceeding the limits of its capacity, – constant protection and, if possible, regeneration of biological diversity at four levels: landscape level, ecosystem level, species level, genetic level, – creating conditions for economic entities for fair competition in access to limited resources and opportunities to discharge pollution, – socializing decision-making processes, especially regarding local natural environment, – providing human individuals with a sense of ecological security, understood as the creation of favorable conditions for physical, mental and social (with local bonds) health.
37. B. Poskrobko (1997)	Three dimensions of sustainable development: – socio-philosophical idea, which shows the need to change previous values shaping Euro-Atlantic civilization and to harmonize the relations between economic and non-economic human activity and the natural environment, as well as to introduce non-antagonistic relations between various systems and social groups,

(Continued)

Table 6. (*Continued*).

	– this idea refers to the modern direction of economic development, based on the modern way of organization and management over economic units, whose technological processes are environmentally-friendly and ensure security and comfort to the people, – it points at a new direction in scientific research, integrating knowledge from various departments in order to learn more about the relations in the macrosystem: society—economy—environment.
38. A.A. Rassafi, H. Poorzahedy, M. Vaziri (2006)	A definition, which places the discussion in the context of general systems theory. A system is sustainable if it is durable (in time), its structure is organized (non-chaotic), subject to limitations (standards, levels of tolerance, restrictions, etc.) and is based on components.
39. M.R. Redclift (1987)	Sustainable development is not devoid of analytical content; it is more than a search for a compromise between natural environment and the pursuit for economic growth.
40. M.R. Redclift (1987)	It is such a development, within which sustainability is both structural and natural. As long as there are impoverished people, there can be no sustainability.
41. I. Sachs (1995)	Development directed at harmonizing social and economic goals with ecologically reasonable management.
42. Z. Sadowski (1998)	Development, within which the increasing ecological costs of development, threatening the living conditions of future generations in a global scale, are eliminated due to conscious directing of human activity towards environmental protection.
43. V. Shiva (1989)	Includes the context of sustainable development in the aspect of sustainability in nature, which can only be achieved by means of maintaining the integrity, cycles and rhythms of natural processes.
44. Stappen (2006)	It is a development that satisfies the basic needs of all people, at the same time caring for the protection, conservation and integrity of Earth's ecological systems, without the risk that the needs of future generations would not be satisfied, or that the limits of Earth's capacity would be exceeded.
45. D. Szaniawska (1995)	Sustainable development is based on a compromise between economic and civilizational development and on the preservation of the environment for present and future generations in as good state as possible.
46. J. Szleszynski (1997)	A strategy of improving the quality of life within the limits of the environment's capacity.
47. European Union: The 5th Environmental Programme (1993)	Sustainable development reflects the policy and strategy of constant economic and social development without harming the environment and the natural resources.

(*Continued*)

Table 6. (*Continued*).

48. FAO (1990)	Sustainable development is management of natural resources, protection thereof and forcing technological and institutional changes in order to meet the needs of present and future generations. Sustainable development is harmless to the environment, technically correct, economically efficient and socially acceptable.
49. A. Zufal (1986)	It is conducting any economic activity in harmony with nature in such a way as not to cause irreversible changes in nature.
50. T. Zylicz	It should satisfy the current needs with no harm to the opportunity of future generations to satisfy their needs. Assuming the needs and the capital productivity are stable, this implicates the necessity to maintain this capital at a non-decreasing level.

The definitions presented above show that sustainable development is not just another economic program. It is also not another concept of environmental protection or nature conservation. It is an attempt at formulating a program integrating various planes of human activity (previously considered separately), based—as will be pointed out—on a moral reflection regarding human responsibility for nature. This integration means seeking to reach an order in all those planes (Borys, 1998, 2005), and a clear distinction of individual problem groups of sustainable development is an important task in this context.

2 HIERARCHY OF PLANES

At present, the majority of publications within the discussed issues only recognize ecological, social and economic aspects (Harris et al., 2001; Holmberg, 1992; Reed, 1997). In my opinion, however, the list should be expanded to include the following viewpoints (Pawłowski, 2004, 2006, 2007, 2008, 2010):

- Ethical plane (the issue of human responsibility for nature).
- Ecological plane (protection of both natural and artificial environment, also includes spatial planning).
- Social plane (social environment may also degrade).
- Economic plane (taxes, subsidies and other economic instruments).
- Technical plane (new technologies, saving resources).
- Legal plane (environmental law).
- Political plane (formulating strategies of sustainable development, introduction and control thereof).

These dimensions can allow the construction of a structure with a hierarchical system of three levels—see Table 7.

The first level, which is the foundation for the others, is an ethical reflection. It is the most important level, determining human activity.

Table 7.   Hierarchy of the planes of sustainable development (Author's own work).

Level I		Ethical plane	
Level II	Ecological plane	Social plane	Economic plane
Level III	Technical plane	Legal plane	Political plane

Level two covers ecological, social and economic issues, all treated as equally important.

The third level is an analysis of technical, legal and political issues. It is as important as level II, however it covers more detailed problem areas.

The traditional discussion about sustainable development concentrates on the second level. It would be incomplete, however, if not rooted in ethics (level one). Without level three, on the other hand, actual practical detailed solutions may be excluded.

It needs to be pointed out that the planes in question are interpenetrative, which makes it hard to discuss problems solely characteristic for one of them only. Even in the case of fulfilling mankind's nonmaterial needs, we cannot avoid associations with the environment. This results from the biological principles of the functioning of human body, which is in constant need of food and therefore interacts with the environment in this sense at least (Littig & Griesler, 2005), although human beings do not have only material needs. Therefore, a number of issues raised in this book are analyzed in ecological, social and economic aspects equally.

The interpenetration of planes in sustainable development shows the necessity to integrate them within a wider vision (Michael et al., 2003; Charmont, 2000; Wilson, 2001). Not only do we speak many national languages, but have also developed numerous scientific disciplines within them, all of which are subject to a progressive specialization, has its own language (which is often unintelligible to representatives of other disciplines). This situation may only be changed by a wider synthesis. At first, such tendencies were observed in physics. The basis of Maxwell's new thermodynamics (already in the 19th century) and Einstein's theory of relativity and relative mechanics (in the 20th century) gave way to the discoveries of nuclear physics, leading to the development of non-relativistic quantum mechanics of the micro-world. This, in turn, entailed a factual integration of physics and chemistry (Legocki, 2006).

In the case of sustainable development, the range of knowledge subject to the expected integration is much larger.

The area of interest within natural sciences coincides with Haeckel's classical definition of ecology, and hence regards the study of nature and the interactions between living organisms and their environment.

Within technical sciences, emphasis is placed on the environment's parameters and admissible emission standards, correlated with actual capabilities determined by technological development.

Within the humanities, environmental issues emerged quite recently. However, since the majority of changes going on in nature are caused by human activity, the aspect of responsibility for them (which is a question of ethics) seems very significant.

Before discussing detailed problem groups related to the idea of sustainable development, however, let us try to point out some general rules accepted during the discussions on this development.

3 THE PRINCIPLES OF SUSTAINABLE DEVELOPMENT

Many normative rules were formulated on sustainable development, which form a set of constituted characteristics of the concept (Borys, 2003; Skowronski, 2006; Moffat, 1996). Some of them raise detailed issues (such as "Principles for Responsible Business", presented within the discussion on the economic plane); some refer to general questions associated with sustainable development.

One of the more significant proposals was the concept of the 'conserver society'. It was formulated in 1976 by Canadian scientists from the GAMMA Institute in Montreal: K. Valaskakis, P.S. Sindell, J.G. Smith and J. Fitzpactrick-Martin. In search of new ways to further human development, five possible scenarios were discussed. The first two were derived from the present state and were described with the slogans (Valaskakis et al., 1981):

- to produce more, to consume more,
- to produce less, to consume more.

The first of these two options preserves the *status quo*; the second had been described as the 'wasteful society' or 'anti-conserver society'.

Alternative models were also presented in the work that may be summarized with the short rules:

- to produce more, to consume less,
- to produce the same, to consume less,
- to consume less, to produce less and to produce something else.

The first rule assumes further scientific and technological progress. New technologies may allow for increased production with less material consumption and the same quality of the end product. The second rule enhances this approach with the aspect of reducing consumption. The third rule—the most sophisticated one—is a postulate for seeking alternatives to modern products and production processes that would be less harmful to the environment.

The proposals presented above place the issue of mankind's development directly next to the consumption of natural resources. The scenarios, however, were prepared with wider assumptions in mind, which also included philosophical conditions. These assumptions are listed in Table 8.

It is no coincidence that the first two rules have an estimative and ethical character. Unlike many other programs, the authors were aware that without a proper determination of the human-nature relation, also in terms of the hierarchy of ethical values, the proposed program may turn out fictitious.

Eventually, the proposal of the 'conserver society' was not introduced, but nevertheless turned out to be one of the crucial inspirations during the formulation of the later concept of sustainable development.

Among other initiatives, the rules formulated in the "Rio Declaration" of 1992 are of great significance—see Table 9.

Table 8. The assumptions of the concept of a 'conserver society' (Valaskakis et al., 1981).

No.	Assumption
1.	The goal is human self-realization through life in harmony with nature. In this context, the tasks, such as economical resource management, are only a means, not a primary goal.
2.	The happiness of people depends on reaching an equilibrium between their needs and the available (material and spiritual) goods, which can help satisfy those needs.
3.	Production processes disallow saving on the use of all materials. Therefore, a hierarchy should be established, deciding over which resources are the most valuable.
4.	Various selection criteria may help us in this choice: – economic criterion (the resources that even now are the least available, are also the most expensive), – ecological criterion (those materials, which are crucial to the survival of a particular ecosystem, are the most valuable), – axiological criterion (in this case, a reference to the adopted system of values takes place). A conserver society makes the choice by combining the three criteria.
5.	A number of types of conserver societies may be distinguished.
6.	Each type of such a society must take into account the local (environmental and cultural) conditions.
7.	The main task of a conserver society is to seek to reduce the economic inequality (both at the national and at the international level).

Table 9. A description of the principles of "Rio Declaration on Environment and Development", adopted during the UN Earth Summit in June 1992 (UN, 1992).

Principle	Content
Preamble	Confirms the significance of the "Stockholm Declaration" of 1972, though suggesting the need to enhance it. Establishing a new and equitable global partnership was set as the goal of the "Rio Declaration", through setting up new forms of cooperation at the level of states, nations as well as basic social groups. This partnership must be based on international agreements, which would help to maintain the integrity of the global system of environment and development.
Principle 1	Provides the special role of human beings in the process of introducing sustainable development. It also confirms the people's right to a healthy and productive life in harmony with nature.
Principle 2	Grants the rights to countries, to use their natural resources, provided that such activity does not cause harm to the environment of other states.
Principle 3	States that the right for development must be fulfilled in such a way as to ensure a fair compilation of developmental and environmental needs of present and future generations alike.
Principle 4	Underlines that environmental protection shall constitute an integral part of the process of sustainable development, which cannot be achieved without it.

(Continued)

Table 9. (*Continued*).

Principle 5	Demands eradicating all poverty and states that this is an indispensable requirement of sustainable development. All states and all people shall cooperate in this task.
Principle 6	Notes that actions for environmental protection and development should include the interests of all countries, particularly the poorest.
Principle 7	Confirms the joint responsibility of all states for degradation of the environment. All countries should cooperate in order to protect the environment. Moreover, this principle stresses the fact that responsibility for the Earth is variable, since the impact that individual countries have on the global ecosystem is also variable.
Principle 8	Points to the relation between sustainable development and the issue of the quality of life. The barriers that need to be overcome on the way towards achieving a higher level of living standard are: excessive consumption, unbalanced production systems and excessive human population growth.
Principle 9	Shows the significance of scientific and technological transfer for creating sustainable development.
Principle 10	Imposes an obligation on public authorities to provide every individual with access to information regarding the environment (as well as to legal and administrative means). It is also a postulate to include societies and concerned citizens into decision-making processes, in as wide a range as possible.
Principle 11	Demands enacting effective environmental legislation. The standards set in it must not cause unjust economic costs in other states (especially in the developing countries).
Principle 12	Notes the necessity for cooperation between individual states in order to promote a supportive and open international economic system, which would lead both to economic growth and better solutions to the issues regarding environmental degradation.
Principle 13	Raises the issue of both international and national legislation in the context of responsibility and compensation for the victims of environmental degradation.
Principle 14	Considers preventing the transfer of substances harmful to the environment and to human health to other countries.
Principle 15	Postulates the adoption of a precautionary approach to environmental protection, especially in the aspect of actions, which could entail serious (or even irreversible) damage. The lack of scientific certainty in this case is not a sufficient foundation for delaying preventive action.
Principle 16	Notes that national authorities should endeavor to promote internal means of obtaining necessary funds, regarding the cost of preventing environmental degradation. The 'polluter pays' principle was also addressed.
Principle 17	Demands that competent national authorities perform impact assessments on the environment in the case of actions which might have a significant, harmful influence on the environment.
Principle 18	Demands that, in case of a disaster (or other emergency), individual countries immediately provided information to other states which also might suffer from these events.

(*Continued*)

Table 9. (*Continued*).

Principle 19	Raises the issue of the obligation to provide information (even at early stages) on actions, whose negative effects might cross the borders of the state taking these actions.
Principle 20	Affirms that women have a vital role in environmental management and development.
Principle 21	Draws attention to young people, whose creativity and ideals may play a significant role in the creation of a global partnership.
Principle 22	Postulates the necessity to support the identity and culture of local communities, whose knowledge and tradition have an impact on environmental management and development.
Principle 23	Indicates that the environment should be protected in all states, regardless of political considerations.
Principle 24	Points out that acts of war always lead to the destruction of the environment and are impossible to reconcile with sustainable development.
Principle 25	Says, that peace, development and environmental protection are interdependent and indivisible.
Principle 26	Stresses the fact that the states should solve environmental issues peacefully, based on the "United Nations Charter".
Principle 27	Contains a general postulate of the necessity of international cooperation in the fulfillment of all principles embodied in the Declaration and such development of international law, which would serve the implementation of sustainable development.

First of all, this is so because they were announced within the UN, which is the same organization that proposed the now commonly accepted principle of sustainable development in 1987.

Secondly—although it was prepared at an early stage of formulating this concept—it is still a reference point for further strategies and studies. Sufficient to say that, when the next Earth Summit took place in Johannesburg in 2002, the next Declaration already included the will to further implement sustainable development in point 1, whereas point 8 specified that the principles from Rio are in force (Earthsummit2002.org, 2002).

"The Rio Declaration" placed the discussion in the context of strengthening the wide international cooperation within the UN. It is the right approach, since global environmental issues are impossible to solve within individual states or even groups of states. This cooperation must not only include accepting common legal solutions, but also support the free transfer of science and technology, which is likely to prove crucial in the implementation of sustainable development.

Further, the Declaration includes some ethical standpoints.

Principle 1 points at the special role of human beings in implementing the discussed way of development.

Principle 3 is a repetition of the general formulation of sustainable development from the "Our Common Future" report, hence confirming the importance of the principle of equity within and between generations.

The document also introduced the notion of responsibility in the international (Principle 7) and the individual aspects (Principle 8, the issue of consumerism as well as the more general notion of the quality of life).

50

"The Rio Declaration", though crucial, does not fully exploit the set of normative laws, which were proposed regarding the implementation of sustainable development. Let us examine a selection of three more proposals:

- "The Earth Charter" of 2000 (four problem groups and sixteen principles).
- "The Dutch paradigms of Sustainable Development" of 1994 (four paradigms).
- "The Principles of the Ecological Era" by S. Kozlowski (twelve principles).

Table 10 summarizes the discussion on the "Earth Charter", supervised by an independent commission (World Charter Commission) and accepted in its final form by many important organizations, including the UN.

Table 10. "The Earth Charter" of 2000 (Charter, 2010).

Group	Principles
Preamble	"The Earth Charter" refers to threats to diversity of life forms and human cultures, which should be placed under protection. The goal is to create a responsible and sustainable society now, for the good of future generations.
Respect for and protection of life and its biodiversity	Respect for the Earth and all life. All living beings have an inherent value, regardless of their usefulness to humans. Caring for the Earth, increasing knowledge and human strength entail the necessity to take responsibility for preserving common good. Building democratic societies, which are (socially and economically) just, engaged, sustainable and peaceful. Preservation of the Earth's beauty.
Protection and restoration of the integrity of Earth's ecological systems	Protecting the integrity of Earth's ecosystems, with special care for biodiversity and natural life-supporting processes. Preventing damage is the best way to protect the environment. When knowledge is limited, a preventive approach should be introduced and we should refrain from potentially dangerous activities. Introduction of production and consumption patterns, which do not threaten Earth's regeneration capabilities and ensure human rights as well as the prosperity of societies. Support for research on ecological sustainability and introduction of free international exchange of the necessary knowledge and technology.
Economic and social equity	Fight against poverty as an ethical, social and environmental imperative. Ensuring that institutions and actions of economic character support the human development at all levels in a fair and sustainable way. Confirmation that just equality is a fundamental condition for sustainable development. Equal share of all people (regardless of race, sex, religion) in the access to natural resources and healthy environment.
Democracy, violence prevention, promotion of tolerance and peace	Strengthening democratic processes, which give people an opportunity to effectively participate in decision making in all sectors of the society. Developing and introducing values and skills, which promote sustainable models of life and protect the environment. Treating all creatures with care and protecting them against cruelty and unjustified destruction. Promoting the culture of tolerance and peace, devoid of violence.

The principles of the Charter are fundamentally compliant with "The Rio Declaration". However, it is worth highlight Principle 7, which includes the important term 'ethical imperative'. What is interesting is that it has only been placed in the context of the fight against poverty.

The paradigms of sustainable development proposed by the Netherlands' Scientific Council for Governmental Strategies of 1994 have a completely different character—see Table 11. These concentrated on the issue of mankind's negative impact on the environment and are based on the necessity to change our traditional behavior. The paradigm of management should be marked—it is an attempt to answer the question: What action must be taken if the postulate of reducing the level of consumption turns out impossible to achieve? The authors of the discussed proposal have faith in new technologies (i.e. material-saving technologies).

There are other ways of reasoning regarding the principles of sustainable development, like the general principles of the ecologic era, discussed from the point of view of the conditions of eco-development—see Table 12. They were presented by Stefan Kozlowski, soon after the end of the UN Earth Summit of 1992.

These principles concentrate on environmental aspects, among which economic, social and political conditions are mentioned as well. Rule 1 should be emphasized, which gives priority to one's spiritual development. It is also worth noting that the concept of sustainable development as such was only placed on the fourth place. This confirms the thesis that eco-development is a narrower notion than sustainable development.

The estimation of such sets of principles is still problematic. They have—at least indirectly—ethical (or, more generally, philosophical) character, since they regard various obligations of mankind. They have not been, however, prepared by philosophers, therefore these aspects have not been pronounced properly. Terms like 'shall' or 'should be' are not enough to speak of ethics or philosophy. All these sets of principles (including the principles from Rio) are—essentially—a collection of imperatives that are unclassifiable and unverifiable as a whole. Moreover, even those arranged next to one another are clearly formulated at different levels of content, and—compared to other sets of principles—are even sometimes in conflict with one another. Still, these rules have an important role in the discussion on sustainable development, forming a particular 'background' to the problems.

Table 11. Paradigms of sustainable development proposed by the Netherlands' Scientific Council for Governmental Strategies in 1994 (Sörlin, 1997).

Name	Content
Paradigm of utility	Human actions have an impact on the environment, although they can be neutralized.
Paradigm of protection	The environment's capacity is limited, people need to adjust to it and reduce their level of consumption.
Paradigm of management	If a reduction in the level of consumption is impossible, a solution to the problem may refer to choosing the right technologies.
Paradigm of behavior	Nature is delicate; therefore we must adjust to its rules. However, any changes in our behavior must occur before the extent of environmental is too massive to reverse.

Table 12.   Principles of the ecologic era (Kozlowski, 1993).

No.	Principle
1.	Accepting that a person's psychological (and not material) development is the main purpose of life.
2.	Determining the impassable level of satisfying human material needs, therefore reversing the negative trend of ever-increasing consumption.
3.	Establishing average fertility at two children per family.
4.	Accepting the concept of sustainable development.
5.	Seeking to protect the Earth's main ecosystems.
6.	Gaining enough knowledge on natural resources management in order to stop the degradation of the environment.
7.	Adopting the concept of a supportive open economic system (the stronger partner would support the weaker one).
8.	Adopting the principle of fair trade, based on optimal distribution of goods in the global scale.
9.	Taxing industrialized countries for the developing countries.
10.	Accepting the obligation to create national ecological policies.
11.	Developing the citizens' rights and activities in order to include them in the decision-making processes.
12.	Creating new pro-ecological organization structures, including e.g. World Ecological Police, World Ecological Tribunal, World Ecological Bank and the non-governmental Earth Council.

4   INDICATORS OF SUSTAINABLE DEVELOPMENT

The principles discussed above may be described with a useful tool in the form of indicators (measures) of sustainable development, referring to the general program of actions—"Agenda 21". Such groups of indicators have been devised by the UN, OECD, or EEA. Table 13 contains a sample list of indicators used in the EU.

Such indicator may be divided into a number of groups distinguished according to a few basic criteria (Borys, 2005):

– The field criterion: to what group of problems (or single problem) do they refer. In general, there are environmental, social, and economic indicators.
– The accuracy criterion: does the indicator actually refer to sustainable development, or measures an unrelated or only seemingly related concept.
– The importance criterion: is the indicator fundamental (crucial; is it directly linked to the goals of the implemented sustainable development policy), or is it only supplementary.
– The criterion of indicator comprehensiveness, defining the scope and type of information which can be obtained with its help. It is also a matter of details: the more detailed the measure, the longer the list of indicators; the more general—the shorter the list, but it also contains less information. In practice, the minuteness of available data is not always an advantage, as a number of features can only be seen after data generalisation.
– The criterion of the level to which the indicator refers; it may be global, international, national, and local (these levels are also mentioned in "Agenda 21").

53

Table 13.   The Sustainable Development Indicators (SDIs) used to monitor the EU Sustainable Development Strategy (Eurostat, 2010).

Theme	Headline Indicator	Value in UE-27 in 2007/2009*
Socio-economic development	Growth rate of GDP (Gross Domestic Product) *per capita.*	20 600 Euro per inhabitant*
Sustainable consumption and production	Resource productivity (GDP divided by domestic material consumption (DMC). DMC measures the total amount of materials directly used by an economy.	1,3
Social inclusion	Percentage of total population at risk of poverty, or social exclusion.	24,5%
Demographic changes	Employment rate of older workers.	54,6% of males and 37,8% of females*
Public health	Healthy life years and life expectancy at birth, by gender.	62,3 for females and 61,5 for males
	Life expectancy.	82,2 for females, and 76,06 for males
Climate change and energy	Greenhouse gas emissions in CO_2 equivalent indexed to 1990, index base year = 100%.	90,5
	Share of renewable energy in gross final energy consumption.	9,7%
Sustainable transport	Energy consumption of transport relative to GDP, so the ratio between the energy consumption of all transport (except of maritime and pipelines) and GDP (chain-linked volumes, at 2000 exchange rates = 100).	95,5
Natural resources	Common bird index, which is an aggregated index integrating the population abundance and the diversity of a selection of common bird species associated with specific habitats (without rare species). It presents farmland species, and all common bird species.	Farmland species: 80,9 Common species: 91,3 (100 = 1990 year)

(Continued)

54

Table 13. (*Continued*).

	Fish catches taken from stocks outside safe biological limits.	21% (data from 2006)
Global partnership	Official development assistance (ODA, consists of grants or loans that are undertaken by the official sector) as share of gross national income.	0,42%*
Good governance	No headline indicator.	–

- Within the level criterion one might also point to universal indicators (e.g. general European indicators) and specific indicators (local ones, related to a specific place e.g. commune or district).
- The time criterion: one distinguishes not only long-term indicators but also those with a shorter time perspective.

All these indicators refer to the commonly accepted definition of sustainable development—the one from the UN report "Our Common Future" of 1987. It refers to the rights of the present and future generations. This definition has been selected in this work as a starting point for the reflection on philosophical conditions of the ethical plane (so level I) of sustainable development and will be the subject of analysis in the next part of the book.

CHAPTER 3

Philosophy, religion and environmental education

Strategies of sustainable development, like any other action program, require rational thought. However, the recommendations such strategies include often concentrate only on technical parameters of the environment and on economic conditions, which, although important, do not provide a complete answer to the fundamental question: why should we behave in one way and not in another? This is because the discussion within them mainly regards levels II and III of sustainable development. There is still level I, however, which consists of philosophy—a discipline which, in its long history, has very frequently asked the question: why? (Graaf et al., 1996).

1 ECO-PHILOSOPHY AND THE ETHICAL PLANE OF SUSTAINABLE DEVELOPMENT

Alexander King, co-founder of the Club of Rome, once said in the context of contemporary environmental challenges: ecology needs philosophy. This short statement bears profound content. After all is said and done, philosophical reflections seek to formulate a relatively consistent system of views, reaching a state of cohesion of the symbolic universe, which could be of significant help in arranging the generally disordered image of the modern world. In this context, the concept of sustainable development, although not a philosophy *per se* (the principle of sustainable development is merely an ethical imperative), fits into the modern philosophical, and more specifically: eco-philosophical discussion. Philosophy refers to all that exists, whereas eco-philosophy only refers to the determinants of the human-nature relation. How can we describe eco-philosophy more precisely?

Jozef M. Dolega suggests that it is a systemic approach to the philosophical issues of socio-natural environment, hence regarding the Earth's ecosystem and its surroundings (Dolega, 2006). Zdzislawa Piatek adds that it is a new philosophy of action. One that wants to establish a human world, seeking human specificity in their symbiosis with nature and not in the contradictions between them (Piatek, 2007). Eco-philosophy deals with, general philosophical, anthropological, axiological and educational issues—see Table 14.

Eco-philosophy in this context has important descriptive and explanatory functions: first of all, it tries to determine the humans' place in nature, their relations with the environment; second, it indicates the system of values necessary to make people more sensitive to the nature.

Table 14. The basic substantive issues of eco-philosophy (Dolega, 2006).

Problematic group	Description
General philosophical	The essence and nature of the natural as well as the social environment; changes, characteristics (quantitative and qualitative), mutual relations (degradation of the environment on one hand, its impact on humans on the other). Searching for the philosophical foundations of environmental protection.
Anthropological	A reflection regarding structural elements of the anthroposphere: science, technology, art and religion.
Axiological	This group regards valuation. Life is said to be the utmost (although not absolute) value, whereas human health is said to be the core value and a common good.
Educational	This is the shaping of ecological awareness at all levels: from children to adult education.

The earliest more elaborate eco-philosophical reflections appeared in the mid-20th century (the works of Aldo Leopold), although they remained outside the interest of the majority of philosophers. Changes in overall approach took place in the 60's and 70's (Tyburski, 2008), when the discussion included, inter alia, the considerations of Arne Naess (founder of deep ecology), Hans Jonas (ethics of responsibility, analyzed further in this work) and in Poland e.g. Henryk Skolimowski (the creator of the term 'eco-philosophy').

Despite vague philosophical considerations, or even—so far—lack of fixed epistemology and methodology, eco-philosophy advances dynamically, with an eco-philosophical reflection regarding the ecological crisis at its source.

The very notion of 'crisis' derives from Greek *krisis* and means conclusion, abrupt change, as well as catastrophe, overproduction and judgement (Pawłowski, 1999; Kolakowski, 1990). On the other hand, the ecological crisis is often perceived as a state of destruction of the environment's ecological values—in various regions or in the entire biosphere—by humans (Cichy et al., 1988). In the opinion of the present author, this statement is too radical. Although local degradation of the environment, global threats (such as the greenhouse effect) and constant reduction of biodiversity are indisputable facts, the biosphere has not been destroyed. In this aspect, the ecological crisis is more of a warning: a catastrophe may occur. It therefore encompasses a state of insecurity (environmental risk) of a particular system (e.g. an ecosystem). It can still return to its former state, but can also be destroyed or at least lose some of its functionality. This risk refers directly to human activity, whereas a crisis is a subjective means of assessment of a given status, noting that the particular system does not function normally.

Moreover, an ecological crisis is more than just environmental destruction, which is most clearly depicted with the concept of sustainable development. At its source is an incorrect—though historically determined—way of shaping human-human and human-nature relations. It is also one of the fundamental issues of eco-philosophy, whose assessment takes place on the grounds of environmental ethics (Piatek, 1998).

According to the definition by Antoni B. Stepien, ethics is a philosophy of morality, which is a theory of moral values and of moral conduct and deals with norms and valuations regarding human behaviour (Stepien, 1989).

Morality is understood as "a certain (rational) property of any aware and free human act and the doer or performer of this act, a human being. (...) Either it is compatible with the doer's nature and leads to achieving their ultimate goal (it is then morally good), or it is incompatible with the doer's nature and hinders achieving their ultimate goal (it is then morally wrong), or it is incompatible with the nature of the doer and does not hinder achieving their ultimate goal (it is then morally neutral)" (Stepien, 1989).

Whereas environmental ethics, as Wlodzimierz Tyburski points out, deal with moral relations between an individual, social groups and entire communities and the humans' natural environment (Tyburski, 2004; Bonenberg, 1992). To express all this differently, ethics becomes environmental, when it takes up the reflection on the human-nature relations. W. Szymanski adds: "it should be shaped as a practical moral discipline, which could provide a total revaluation of our human behaviour toward the world of nature. (...) Ecological ethics is presented as a model of thinking, and hence as a set of conceptual formulations, which would serve practical applications for (...) resolving all problems regarding the relations between humans and natural environment in the light of the notions, values and moral standards as well as in valuation categories of moral good and wrong" (Szymanski, 2004). In other words: we need a development that would not only be economically and ecologically sustainable but morally sustainable as well (Stenmarck, 1997). Moreover, without this feature, sustainability within other planes can turn out impossible. Therefore, the ethical plane of sustainable development (at both theoretical and practical levels) provides a starting point for other planes.

We are actually dealing with two types of relations:

- Planes of sustainable development and their moral assessment.
- Internal interactions between individual planes and their assessment.

The planes discussed in this work include various issues, but are nevertheless closely related. The ecological plane is an example of this: there are more and more protected areas, but the current state of the natural environment is not necessarily improved. Shortcomings in other planes are an obvious cause of this. And vice versa: it is impossible to achieve an improvement in the social plane without improving the natural environment as well.

Paradoxically, this proves that the concept of sustainable development is valid. The time of pure ecology or environmental protection is over. Interdisciplinary programs and activities are required. Just as it is not true that all problems would be resolved by the invisible hand of the market, so it is also not true that they would be resolved by the invisible hand of ecology (Shrader-Frechette, 2006). The planes of sustainable development have much in common from the view of environmental ethics as well. In all cases, the starting point should be a reflection concerning human-human and human-nature relations. It is a question of the rights of both humans and nature and of the obligations directly related to humans and their activity in respect of the environment.

Also, the answer to the question, why should we act in this way and not otherwise, may be manifold (Ryden, 1997):

- Legal answer: because that is the law.
- Social answer: because others do so.
- Psychological answer: because I feel that such behaviour is appropriate.

- Instrumental answer: because it is most effective.
- Ethical answer: because I respect particular values.

We will return to the issue of values, but let us first present the general determinants of environmental ethics, out of which two important problems should be emphasised.

First of all, the global conditions of human life, or the possibility of its survival, have scarcely been discussed in the long history of philosophical reflection. In earlier times humans did not possess sufficient technological wherewithal to destroy the entire biosphere and ecological disasters usually only occurred in a purely local aspect. This has changed, however.

Second, the 'classical' reference of moral assessments only to direct results of human activity seems insufficient as some consequences of human activity (especially technical activity) may manifest themselves years after the opportunities for repair have been severely impeded or even completely eliminated. This also regards humans themselves, particularly in the aspect of genetic changes (usually inherited by future generations) associated with environmental pollution.

In fact, as Andrzej Kiepas notes, it is an issue of the three opacities (hazy issues), mainly regarding human technical activity (Kiepas, 2006):

- opacity of effects: we are unable to foresee all effects of human behaviour,
- opacity of intentions: clear identification of the perpetrator is often impossible in random events,
- opacity of causation: what we are currently dealing with is not a cause-effect relation, but a network of links between phenomena distant from one another in both time and space.

These opacities are a major problem for the representatives of technical sciences and their moral assessment is an important ethical challenge, since there are several various positions of environmental ethics.

The basic criterion of division in environmental ethics is the list of objects with a moral status, which can be distinguished in the human-nature relation. According to this, anthropocentric and non-anthropocentric approaches are highlighted. The former assumes human superiority over nature—exclusionism; the latter assumes a partnership relation—inclusionism (Slipko, 1992).

From the point of view of environmental ethics, anthropocentrism points at the benefits humans could enjoy from taking responsibility towards nature. This direction divides into two positions: a strong one and a weak one.

- The strong position assumes the environment is merely a means to the ultimate goal, which is satisfying human needs (Norton 1993). It can therefore be described as extremely utilitarian.
- In terms of the weak position, known also as moderate anthropocentrism, the meaning of nature as a necessary factor to human existence is being highlighted. Obligations towards future generations, who should live in a world as good as the one we live in, are also pointed out (Norton, 1993; Murdy, 1993). This is the plane, within which the Christian vision of nature (sometimes described as theocentricism—as opposed to the secular anthropocentrism) has been formulated. This is also the standpoint of Hans Jonas, whose works will be discussed below.

60

A human being is the only moral entity within anthropocentrism. Consequently, it is thought that nature—either the whole of it or its individual elements—cannot be given moral status. At the same time—a point needing emphasis—it is difficult to imagine an activity that would cause irreparable harm to the environment or to an ecosystem, without threatening human beings (Chrader-Frachette, 1996).

Two positions can be distinguished in the non-anthropocentric approach as well.

- Biocentrism, which grants the right to existence to all living creatures (humans and animals alike), giving them moral status. In this context, not only humans, but also other living things are given an internal value. The major reason is the hypothesis that, since animals can suffer, they should be protected. This concept was expanded by e.g. T. Regan and A. Schweitzer.
- Ecocentrism (also described as a holistic, systemic foundation). Its supporters recognize the value of whole entities created by individuals, such as an ecosystem, or even the biosphere. This is the standpoint of A. Leopold and J. Lovelock.

Table 15 presents basic eco-philosophical concepts.

Paradoxically, all eco-philosophical concepts, however diversely classified, are essentially anthropocentric. After all, only a human being can comprehend 'morality' or the meaning of 'human responsibility for nature'. Moreover, animals fit into the realization of only one informational-biological program. In case of humans, there are more programs: cultural, social, political and moral. This is a significant difference. We may agree that humans are part of nature, but their abilities exceed its limits nevertheless. We cannot perceive the world from other than a human perspective, even if we deny the fact. This is a specific 'trap of anthropocentrism'.

Still, humans created neither nature nor the rights governing it. Moreover, by degrading the world of nature, they threaten themselves, since they are unable to survive without an environment with certain characteristics; hence the imperative of environmental protection. But, not only that, in the ethical perspective, even if only interpersonal relations had a moral character, they would still indirectly refer to nature as well, because of the very fact of the necessity of the environment to humans.

Moreover, crossing the borders between anthropocentric (in moderate aspect) and non-anthropocentric positions in ethics is not as improbable as it may seem. Zdzislawa Piatek notes: "if the key anthropocentric interest—namely, avoiding self-destruction—can be achieved by cooperation with evolutionary processes, then biocentrism turns out to be the reverse side of anthropocentrism. Contrasting the arrogant anthropocentrism [which is the strong version of this position—Author's note] against biocentrism takes place when humans place themselves above nature, or claim to be anti-natural beings. (…) Human species, as one of the millions within the Earth's biosphere, (…) is threatened by self-destruction, similarly to other species. However, by knowingly determining their relations with the environment, they can seek to avoid self-destruction. (…) From the human point of view, in the light of their will to avoid self-destruction, it is necessary to respect nature because of what it is in itself" (Piatek, 2007, 2008).

Table 15.   Chosen eco-philosophers and theirs thesis (Author's own work).

The author	Main thesis
Aldo Leopold	Leopold's main thesis is that a thing is good, when it leads to maintaining the integrity, stability and beauty of the biotic community. If it doesn't, it is wrong. This concept is illustrated with the so-called 'land pyramid', referring to ecological principles governing terrestrial ecosystems (Leopold, 1949).
Arne Naess	Naess introduced the term 'deep ecology' in reference to the deep, spiritual approach toward nature (ecosophy), in opposition to traditional ecology targeted at humans. Naess (1993) points at two intuitive beliefs (primary intuitions): self-realization (internal, spiritual development) and biocentric equality (all elements of the biosphere have the right to live, develop and to self-realization).
Henryk Skolimowski	One of the basic questions among his works is: how to overcome the current human crisis? Skolimowski (2003, 2007) suggests that the crisis has not been caused by humans themselves, but by civilization and more specifically—by the 'unholy trinity': objective science (which ceases to 'see and understand'), destructive technology and the manipulative and mechanistic human mind (which leads not to wisdom, but to manipulation). As a result, humans broke off from nature, of which they are a part, and there is disarray in their way of thinking, which leads to fear and aggression. The crisis is possible to resolve, however, by creating New Normative Knowledge (science serving humans and life), New Technology (whose imperative must be to avoid doing damage) and a new socio-political system (described by Skolimowski as the Cosmocracy).
Tom Regan	His concept regards the 'subjects-of-a-life', which have a psychophysical identity. According to Regan, this condition is most obviously met by adult mammals (Regan, 1993).
Albert Schweitzer	His concept of 'reverence for life' is that sustaining life (which has an autotelic value) is good, whereas destroying it is wrong. This principle regards all life forms (Schweitzer, 1993).
James Lovelock	The author of the Gaia hypothesis, according to which the biosphere is a self-regulating system with the capacity to keep our planet healthy by controlling the physical and chemical environment (Lovelock, 1989, 1995). Sometimes this entity is presented by Lovelock as an immense organism, a living system, which is controversial (how to define life on a planetary scale?) and in later works is treated as a metaphor.
Hans Jonas	The author of the imperative of responsibility for the future generations, which refers both to securing the future for humans, which Jonas deems necessary and to the future conditions and the quality of life (Jonas, 1984).

There are other routes to an 'eco-philosophical agreement'. The Ecosophy T. system (named after a Norwegian mountain Tvergastein), proposed by the founder of deep ecology, Arne Naess (Naess, 1993), is worth mentioning. It is not so much about the full agreement on all detailed solutions suggested by this author, as about the unconventional attempt at proving that some conclusions may be reached from a variety of—contradictory at times—positions. Such an approach fits into the universalistic trend in philosophy. In this context, the division between anthropocentrists and non-anthropocentrists loses its relevance. Moreover, from nature's point of view, it is probably unimportant, whether someone protects it because he or she is an anthropocentrist or e.g. a biocentrist. What is important is the very fact of protection.

How should sustainable development be described in this context?

It is worth remembering the basic principle of this development, introduced by the UN in 1987, as a result of the proceedings of WCED (World Commission on Environment and Development). According to this, "sustainable development is a development that meets the needs of the present generation without compromising the ability of future generations to meet their own needs" (WCED, 1987).

This principle is an imperative, which, although not formulated by philosophers (it was a result of a political compromise), still contains an ethical message, and one with a universal character, regarding global developmental processes. It is an optimistic message, which suggests that realization of this principle is possible and would help avoid humanity's self-destruction (Piatek, 2007).

However, certain philosophical assumptions should be indicated in this principle (Papuzinski, 2004, 2008):

- Ontological assumption: humans occupy a privileged place in the universe; nature is their environment and is discussed through the perspective of humans (their needs, their existence, their welfare). It is a form of moderate anthropocentrism.
- Anthropological assumption: humans are social beings and therefore they self-realize in relations with other people. It should also be emphasised that they are natural beings as well, since it is not possible for them to live without the environment. There is, therefore, a specific co-evolution taking place between two important dimensions of human nature: the biological dimension and the socio-cultural dimension.
- Axiological assumption: responsibility for other people (the present and future generations) is the ultimate value.
- Historiosophical assumption: recognizing that the process of human development is historical.

To summarize, the principle of sustainable development is anthropocentric (moderate anthropocentrism), since it regards the human environment and the quality of human life.

On the other hand, in reference to the environment it is also a call for balancing the odds between humans and nature, which is yet another version of the principle of self-regulation, well-known from the works of Ernst Schumacher (Schumacher, 1973).

In this context, the principle of sustainable development points to four primary objectives (Borys, 1999; Liu, 2001):

- Sustaining the quality of the environment.
- Human security.
- Human welfare (in the material context and—through social context—also in the spiritual one).

63

- Equity (intra- and intergenerational, but also at the local, regional and global level).

Connecting it with the proposition of hierarchy of planes of sustainable development, it should be noted that the very principle in question belongs to level I, but its realization refers to level II (ecological plane—the environment; social plane—security, equity; social and economical plane—welfare) and—through level II—to detailed issues in level III.

Moreover, the term 'meet the needs', used in the definition, requires a comment (Wittig & Griesler, 2005). It should be noted that the term refers to the basic human needs, shared by all people, not just those living in the rich North.

When speaking of needs, it is worth referring to the divisions introduced in psychology.

One of the most commonly known is the Abraham Maslow's model, which recognizes five groups of needs, forming a hierarchical arrangement, which means that higher-order needs can only be met after lower-order needs have also been met. The division into these groups, starting from the most elementary ones, is as follows (Maslow, 1954):

- physiological needs (e.g. the need to eat, to drink, to avoid pain),
- safety needs (e.g. the need for care, support, peace),
- needs of love and belonging (e.g. the need for relationship, the need to be loved),
- needs for prestige and esteem (e.g. the need to confirm one's own self-respect),
- needs for self-realisation (the need to have goals, strive to develop one's own capabilities).

Apart from the above hierarchy, Maslow also pointed to cognitive needs (need for knowledge, understanding) and aesthetic needs (need for harmony and beauty).

Moreover, M. Jarosz (Bielecki, 1986) enlarge on these groups by making a distinction between primary (organic, biological) and secondary needs (psychological and social, resulting from the impact of the environment on humans).

In the case of sustainable development, some basic biological needs arise that should be absolutely satisfied—and only the secondary step includes the issue of non-material needs, such as culture (Haq, 1995).

Unfortunately, in a modern society, satisfaction of the basic needs leads not so much to spiritual development, as to further material consumption. Literature has even evolved the term 'consumer society', within which consumption of goods is the primary method of achieving happiness and a goal in itself (Brown & Cameron, 2000; Ropke, 1999; Rees, 1999; Clark, 2007; Wilk, 2006). Although ecological content can be found in the media, even if only in commercials, this does not mean that products presented in such a way actually have any ecological properties. After all, the only goal of advertising is to encourage the purchase of particular goods.

Therefore, the model of a 'consumer society' in the discussion on sustainable development, is opposed to an alternate concept of consumption, which should lead to an extreme reduction in the rate of human consumption of natural resources. However, 'sustainable consumption' cannot be achieved merely by preparing a list of these and other needs or by summarizing the technical opportunities for their satisfaction. The axiological aspect—namely, a reference to values—is also necessary (Laszlo, 2005; Dolega, 2007). This is another important matter taken up by environmental ethics, since the postulate of sustainable development, without supporting it with particular normative values, will remain futile.

One of the proposals was formulated during the Millennium Summit, which took place at the UN in New York on 6–8 September 2000. In the "United Nations Millennium Declaration" that was adopted then, problems resulting from unequal distribution of goods, clearly increasing as a part of the globalization process, were emphasized. Also, the six fundamental values necessary for mankind's proper development in the 21st century were distinguished. These are (UN, 2000):

- Freedom (not only political freedom, but also freedom from famine and poverty).
- Equality (both between nations and sexes).
- Solidarity (of the rich and the poor states for social equity).
- Tolerance (regarding other cultures, beliefs and languages).
- Respect for nature (access to which must be secured for the future generations as well, which is a direct reference to the principle of sustainable development).
- Shared responsibility for peace and proper socio-economic development.

How can the notion of values be set in the perspective of environmental ethics? It should be noted that it has been the object of a wide theoretical discussion. On the basis of the theory of being, value may be assigned to all beings, or only to selected ones, due to human interest. Moreover, the distinction between cognitive values (cognitionism) and 'generated' values (anti-cognitionism) is significant. It may also be suggested that value is subject to experience (subjectivism) or that it is a property of the object (objectivism), or discussed over naturalism (and more specifically—over the so-called 'naturalistic fallacy'). These issues are not raised in the concept of sustainable development; it is a research issue, which still needs consideration. The current discussion is only touching the general definition of values, which can be described as all that, which is valuable, or desirable and is the ultimate goal of human endeavour (Galkowski, 1992).

It can also be said that values (Tyburski, 2004):

- inform us of the primary objectives,
- are evaluative,
- encourage, or even oblige us, to do something,
- foster the mitigation of conflicts between humans and nature.

Moreover, as Z. Hull suggests, human thinking is programmatically axiological. It does not separate itself from evaluations and postulates (as in positivistic concepts), since every actual human being is an integral part of cognition and action (Hull, 2003).

In this context, any discussion on sustainable development should draw attention to universal humanistic values, such as truth (which is the measure of all other values), goodness and beauty (Krapiec, 1990). In this perspective, attention is also drawn to other values, which seem to be of importance to the principle of sustainable development, e.g. life, asceticism (particularly in the context of consumerism), self-discipline, self-realization, contemplation, respect, equity, or safety (Elgin, 2006; Hay, 2005; Grzegorczyk, 1992; Hayward, 1994). Another proposition regarding values such as love, self-respect, respect for others, honesty, truth, non-violence, humility was introduced by Eva M. Kras (Kras, 2010).

If we apply these values to the human-nature relations, we will start a discussion on the ecological values. Let us arrange them in two classifications.

The first one puts the discussion in the context of human effort. Jerzy W. Galkowski (1992) distinguishes the following values:

- Biological values: soil, water, air.
- Economic values: nature's potential and its resources, processed by humans.
- Cultural values: nature understood as an element of our world, but in the spiritual, psychological aspect. It includes e.g. the issue of the beauty of nature and its aesthetic values.
- Civilizational values: all that does not occur in nature, but has been created by humans.
- Political values: nature becomes a carrier for political values in the context of its degradation, which not only poses a threat at the local level, but is also an international issue (political cooperation—or lack thereof—is a crucial element here).
- Moral values: regarding the core of humanity. It is not extrinsic; it does not stand for 'what I have', but for 'who I am'.

Ecological values can be perceived differently. Zbigniew Hull notes three groups of such values:

- Values serving life, sustaining it, as well as ecological balance and human health. This group includes elementary components of the environment, such as: water, air, soil, which—with regard to values—means high quality and lack of pollution.
- Values serving nature (these are conditioned by the eco-philosophical assumption that nature is good), hence recognizing as valuable such actions as: moderate consumption, making rational use of the environment, or taking responsibility for it.
- Values referring to various areas of human activity: universal solidarity, equal access to the resources of the biosphere, intergenerational equity, demographic responsibility, or responsibility for ecological damage.

It is worth considering the category of responsibility, mentioned in various contexts above, which has an important role in the principle of sustainable development. This is also consistent with the conditions of environmental ethics, which also—as Andrzej Kiepas writes—adopts the category of responsibility as its core basis (Kiepas, 1993). Let us take a closer look at its determinants.

According to Roman Ingarden, responsible action is one that means realizing the values, or—more precisely—making the choice between various feasible values (Ingarden, 2001).

Also, according to Hans Jonas, responsibility is conditioned by (Hayward, 1994):

- The existence of a causal power; accepting the fact that actions have an impact on the world (which is the most general condition of responsibility).
- Accepting the fact that these actions are controlled by the doer.
- Assuming that the doer—to some extent, at least—is able to foresee the consequences of their actions (Jonas, 1984).

The agents that are taken into consideration are e.g. the degree of consciousness (whether the doer acted consciously) and the result of a specific action (may be good, bad or neutral).

Such an understanding of responsibility can be moral (natural responsibility) or formal (contracted responsibility).

Natural responsibility set by nature is absolutely mandatory. This includes e.g. parents' responsibility for their children. Its validity is independent of free choice or prior acknowledgement. Formal responsibility, on the other hand, refers to human institutions, legislation and politics (Lipietz, 1995).

In this context, responsibility can be perceived in two ways (Lipietz, 1995):

- Someone is responsible for something that has happened.
- Someone is a responsible being, who understands its obligations.

R. Ingarden expands this reasoning. A person is responsible, because (Ingarden, 2001):

- He/she is responsible for something.
- He/she is taking responsibility for something.
- He/she is held responsible for something.
- He/she acts responsibly.

Numerous detailed situations are possible here, e.g. one can take responsibility for something that, in fact, one is not responsible for, or be responsible for something, but shirk responsibility and not be held responsible.

The concept of sustainable development distinguishes another context of responsibility: responsibility for what only may happen (e.g. the aspect of environmental degradation). It is not a new idea. R. Ingarden had already noted: Responsibility, which only takes place at the very moment of the act would be pointless; it could not play an important part in building up a basis for repairs, which are to be realized in the future (Ingarden, 2001). This aspect of responsibility has been very clearly stressed in the principle of sustainable development, which is focused on the future—by pointing at 'intergenerational solidarity'.

This is a reference to the prerequisite of human purposeful activity, which regards the future as well as the dreams and hopes associated with it. This prerequisite assumes that there will be a future and that humankind will exist in it. The principle of sustainable development is therefore (Sztumski, 2006) both synchronic (since it regards the present generation) and diachronic (since it also goes into the future, successive generations).

However, it is the present generation, which makes the choice regarding the way of our development, the goals and the means of their realization. The problem here is that the future generations do not exist yet and hence cannot make their case (a somewhat similar situation applies to nature, which—though it exists—cannot speak for itself). If so, do they have any rights? This is an important question, since it is hard to discuss moral relations with non-existent entities. It is much easier to put the discussion in the context of the proper shaping of natural and social environment now, which will (indirectly) regard future generations anyway.

It is worth using a comparison. Many works regarding the issues related to environmental degradation claim that it always involves a specific location (or region) but at the same time, is a global issue (e.g. such phenomena as global warming or the ozone hole, both posing a threat to the entire biosphere). In view of the above, discussing particular issues only at the local level is a mistake. It is also suggested that a similar case can be made regarding the relation between problems posing a threat to the present and future generations. At the local level, this would be the parents-children relationship, whereas the global aspect would include the relations between the present and future generations.

One must ask, however: what does the term 'future generations' mean? This, in turn, calls for another question: how can one define 'generation' as such?

Analyzing the conclusions of Dariusz Liszewski (2007), one can point at a 'generation' understood as:

- A population living close in time, shaped by similar experiences.
- In the family context, a generation defines one's place in the family structure (grandparents, parents, children, grandchildren—these are four different generations).
- A generation is also a period, in which children conventionally grow into parents and parents—into grandparents.

Each of these formulations is imprecise, since in this sense, generations overlap in time, which makes it impossible to clearly separate them from one another. The duration of a single generation is also unclear: 30, 50 years, or maybe even more?

What about future generations? In the context of the definitions presented above, children—who are the already existing generation—may be recognized as the next generation. Such a formulation has serious theoretical consequences, particularly for ethics. Although the time horizon in question is narrowed, the reason for human responsibility is also simpler. Still, the issue of long-term consequences of human actions remains a problem. Some of these consequences (like the threat of depletion of resources) regard a much longer period than the one defined by the lifespan of our children. Perhaps this is why the presented proposal did not—so far, at least—gain wide acceptance. More common are the definitions suggesting—as Martin Golding writes—that the future generations are those that could in no case live with the generation, which is to take responsibility (Birnbacher, 2009)—that is the third generation after the present one.

How do intra- and intergenerational relations shape? Table 16. presents some significant distinctions.

The summary of relations shows that there are bonds between particular generations that fit into the general continuity of human development. This continuity shows specific permanent values, such as health, for instance. As Zbigniew Kuderowicz writes, if health has been a value in the past and still remains a value, we may assume it will also remain a value for the generations to come (Kuderowicz, 1993). However, in order to ensure health, numerous conditions must be met, among which the concern for natural environment is very important. Therefore, the table should be expanded by reference to previous generations as well. A responsible attitude towards future generations may be defined as a repayment of obligations to the past generations, the fruits of whose labour we have inherited (Swierczek, 1990). To formulate it even wider: the rights of the future generations can be considered by referring to the general rights of humankind, which depend directly on the status of biological heritage, offered by nature (Bachelet, 1999). This means that nature conservation is also the protection of humans, both in terms of the present and future generations. In the

Table 16. Intergenerational relations in time and in the context of the principle of sustainable development (Czaja, 2005).

Sender generation	Recipient generation	Effect
Parents	Children	A feedback relation, transfer of values, opportunities for direct communication and agreement on decisions or choice of objectives, with parents' predominance.
Parents	Adult children	A feedback relation, transfer of expectations, opportunities for direct communication and agreement on decisions or choice of objectives, with an increasing role of the children and the possibility of an intergenerational conflict (e.g. youth rebellion).
Grandparents	Grandchildren	Limited feedback relation, transfer of values, opportunities for direct communication, limited possibility of agreeing on decisions or choice of objectives, on account of the declining position of the grandparents' generation.
Earlier generation	Further generations (more than the 3rd)	A one-way relation, material and symbolic traces, lack of opportunities for direct communication or agreeing on objectives (decisions).

ethical perspective, what this means is that humans are not only obliged to take responsibility for themselves but for nature as well.

Such an understanding of responsibility makes it worthwhile to also include Hans Jonas's conclusions regarding the ethics of responsibility, presented as a specific and philosophically coherent system in 1974 (Jonas, 1974). The choice of the author is not accidental—his imperative of responsibility turns out to be close to the later principle of sustainable development.

Jonas's reflections were conducted on the basis of philosophy of technology. According to this author, the enormous technical powers currently available to mankind imply the emergence of new moral issues related to the possibility of destroying the biosphere as well as long-term consequences of human activity, both toward other humans and toward the environment. And since we do not know whether life exists anywhere else in the Universe, our responsibility becomes a cosmic responsibility. This implies a qualitative, metaphysical change, a change in the very essence of human behaviour. What is important, the author does not seek to formulate a brand new ethical concept, but instead to complement—as he describes it—'traditional' ethics. This complementation is based on a formulation and on an attempt at validating the 'imperative of responsibility'. This imperative refers to Immanuel Kant's categorical imperative, so "act so that you can will that the maxim of your action be made the principle of a universal law" (Hayward, 1994).

The imperative of responsibility reads as follows (Jonas, 1984):

- "Act so that the effects of your action are compatible with the permanence of genuine human life".
- "Act so that the effects of your action are not destructive of the future possibility of such life".
- "Do not compromise the conditions for an indefinite continuation of humanity on Earth".

- "In your present choices, include the future wholeness of Man among the objects of your will".

The convergence with the later formulated concept of sustainable development can be easily appreciated. The imperative of responsibility, in this understanding, is twofold:

- it refers to securing the future for humans, which Jonas deems necessary,
- it refers to the future conditions and the quality of life.

Moreover, it is based on a specific concept of both humankind and nature.

This concept is a reference to H. Arnedt's and R. Bultman's anthropology, as well as the works of M. Heidegger and E. Husserl or E. Bloch's the principle of hope (Bloch, 1959).

In the human aspect Jonas's position is best described as moderate anthropocentrism. It should be noted that this concept refers not so much to individual people as to the society (aggregate). What matters is the collective 'doer' and collective action (Jonas, 1974, 1984). The driving force of an individual did not increase significantly, compared to the past centuries; however, the driving force of the society—the postulated aggregate—has undoubtedly been multiplied many times.

Speaking of nature, it should be noted that it is not only perceived as the human habitat, but also as the condition of our survival.

Jonas seeks validation for his imperative of responsibility in the autotelic value of existence (not necessarily human; it is about the very existence, not about its form), which is recognized by humans as absolutely mandatory. The author is asking: why is there something rather than nothing (which is a reference to the philosophy of G.W. Leibnitz). He does this convinced that the intrinsic value of life manifests itself in confrontation with nothingness. In other words: Jonas finds the value of what exists in an analysis of the presence of purposefulness in nature (Ferry, 1992), which is focused on sustaining the existence of living world (where the very existence is the value). So, Nature, also having goals, puts values into existence. For every goal achieved brings about particular good things. A goal as such validates itself within the being (Jonas, 1984). This statement means that purposefulness is an ontological property of being. Jonas takes it for an ontological axiom. In this context, life is seen as a confrontation between existence and non-existence; as a confirmation of the good ingrained in it.

The author of the imperative of responsibility claims that although the good present in the world cannot force humans to respect it, it is the basis of an absolutely mandatory responsibility. Although nature has no powers that could threaten it (since the mechanisms in charge of its functioning are characterized by the pursuit for balance) humans do have such powers. Due to the character of these powers, the postulated responsibility should be taken in terms of realizing the potential risk associated with this or that activity (Jonas, 1984; Sztombka, 1991).

However, Jonas's work goes far beyond the validation of the principle of responsibility. It is also an attempt to answer the question, how to act in order to properly fulfil this responsibility. The author suggests accepting two obligations (Jonas, 1984):

- To illustrate the long-term technical consequences of human activity. Comparative futurology is mentioned as a means to fulfil this obligation by comparing all possible scenarios of the course of events.

- Adopting the sense of civilizational fear (so-called 'heuristics of fear'), proportional to the threat.

Since achieving complete knowledge of the future is not possible, futurological forecasts will always be burdened with a certain level of error. Therefore, in order to reduce possible threats, the principle of minimal objectives should be accepted. In other words, humans should not focus on what is technically possible, but on what is necessary. Civilizational fear calls for a particular form of self-control (Sztombka, 1994).

Moreover, out of the many anticipated consequences of human behaviour, we must choose the worst and define remedial action against it. As the aforementioned author writes: the prophecy of doom is to be given greater heed than the prophecy of bliss (Jonas, 1984). It is worth noting that principle 15 of "The Rio Declaration" is framed in a similar vein.

New ideas are more optimistic. Let's mention backcasting—see Figure 1. The basic question is: what do we need to do today to reach that successful outcome? To put it other way, "backcasting is a method in which the future desired conditions are envisioned and steps are then defined to attain those conditions, rather than taking steps that are merely a continuation of present methods extrapolated into the future" (Wikipedia, 2011; Michnowski, 2010).

There are 3 steps of such way of thinking (Natural Step, 2010):

1. Begin with the end in mind.
2. Move backwards from the vision to the present.
3. Move step by step towards the vision.

But how should these proposals be placed in the context of sustainable development?

As already pointed out, Hans Jonas's principle of responsibility is fully compatible with the principle of sustainable development. It is a position of moderate anthropocentrism, based on an original argumentation regarding the autotelic value of existence.

Detailed solutions are more controversial, e.g. a proposal of 'negative' (based on fear) instead of 'positive' ethics (Ciazela, 2006). This leads the author toward

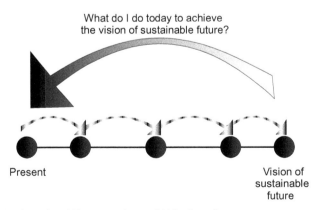

What do I do today to achieve the vision of sustainable future?

Present

Vision of sustainable future

Figure 1.   Backcasting (The natural step, 2010, altered).

71

Table 17.   Declaration "Ethique et spiritualité de l'environment" formulated in Rabat in April 1992 (Zieba, 1996).

No.	Adopted principle
1.	The fate of humankind and nature are inextricably bound together.
2.	Humans are responsible for nature.
3.	Humans should respect the environment, just as they respect their own lives.
4.	Humans bear the moral obligation to wisely and responsibly manage natural resources (biotic and abiotic alike).
5.	Humanity as a whole bears responsibility for nature, as it is the only way to regress the degradation process.
6.	Responsibility obliges humankind to respect a proper scale of values and ontology of human existence, which assumes an alliance of mankind and universal nature.
	The believers take God as reference, the non-believers take common welfare.

the commonly known motto: 'above all, do no harm'. In the case of Jonas this is because of lack of faith in the possibility of humankind's radical transformation. Whereas the principle of sustainable development also does not assume such rapid transformation, instead creating mechanisms (e.g. political, legal, economical), which would also enable significant changes, although introduced gradually and over a longer time range. Also backcasting, is more optimistic that Jonas' heursitics of fear.

These conclusions do not change the general usefulness of the argumentation regarding the principle of responsibility as such in the discussion on sustainable development. Since the adopted strategies include the principle of sustainable development, but do not include its philosophical background, Jonas's eco-philosophical discourse is an interesting suggestion, which points at how this gap can be filled.

On the other hand, in the normative aspect, the chosen path may differ from the one set by the author of the principle of responsibility. It is worthwhile to present an actual proposal, which may also summarize the previous discussion on responsibility—namely, the declaration "Ethique et spiritualité de l'environment" of 1992—see Table 17.

We should stop at the last point of the declaration, because in our search for arguments for the principle of sustainable development, it is worth referring to great religions as well. Their principles have a direct impact on believers and moreover, apart from theological bases, they are also based on general philosophical assumptions.

2   RELIGION AND SUSTAINABLE DEVELOPMENT

Despite their different beliefs, the major religions agree on the issues regarding human relations with the environment, or treating nature as *sacrum* (Gardner, 2006). This was clearly shown at the meeting, which took place in Assisi in 1986. Representatives of five major religions were present: Christianity, Judaism, Islam, Buddhism and Hinduism. A common "Declaration on Nature" was prepared, which included individual declarations motivated by the dogmas of individual religions (Assisi Declaration, 1986). The cooperation on environment conservation was further continued e.g. in 1995 in Windsor, England, or in 2000 in Nepal, when representatives of other religions such as Shintoism and Zoroastrianism had joined the discussion (Alliance of Religion, 2010). We should also mention the even more detailed "Ecological Decalogue", formulated on the Conference of Jews and Christians in Halling Hoog in October 1982—see Table 18.

Religious reflection on the human-nature relation is particularly expanded in Christian ethics, which makes it worthwhile to present this perspective in more detail. First let us remind ourselves that, in line with the long history of ethics, only humans are given the status of moral beings, thus placing them above nature. In this case the human-environment relation becomes part of a wider human-environment-human relation, where the environment merely acts as a mediator.

The adoption of such an approach is often claimed as the main cause of environmental devastation (White, 1967; O'Riordan, 1981). This reasoning is reinforced by the Biblical order, which reads: "Be fruitful (...) and multiply and replenish the earth, and subdue it" (King James Bible, Gen 1:28). But in Christianity, God is the only Ruler of the world and its nature. Making man in the image of God does not separate us from other creatures, nor does it make us masters of the world (Szafranski, 1993). As Pope John Paul II wrote in his encyclical "Sollicitudo rei socialis" of 1987 (later repeated in precisely the same way in "Evangelium vitae" of 1995): "The dominion granted to man by the Creator is not an absolute power, nor can one speak of a freedom to 'use and misuse', or to dispose of things as one pleases. The limitation imposed from the beginning by the Creator himself and expressed symbolically by the prohibition not to 'eat of the fruit of the tree' shows clearly enough that, when it comes to the natural world, we are subject not only to biological laws but also to moral ones, which cannot be violated with impunity."

Table 18. The Ecological Decalogue (Kalinowska, 1992).

No.	Commandment
1.	I am the Lord, your God, who made heaven and earth. Take into account that you are my partner in this creation. Therefore, you shall handle air, water, soil and animals carefully, as if they were your brothers and sisters.
2.	Take into account that while giving you life, I have also given you responsibility, freedom and Earth's limited resources.
3.	You shall not steal from the future. Respect your children and give them opportunity for a long life.
4.	You shall inspire the love of nature in your children.
5.	Consider that humanity may use technology, but it cannot recreate the life that has once been destroyed.
6.	You shall ensure that groups of people form in your village, in your town, in your country, which will engage in preventing ecological disasters.
7.	You shall refrain from using weapons, which cause irreparable damage to the fundamental conditions of life.
8.	You shall exercise self-discipline, even in the small decisions in your life.
9.	Find time in your weekly day of rest—be it Sabbath or Sunday—to live with the world and not just use it.
10.	Consider this: you do not own Earth—you are merely its caretaker.

Furthermore, the biblical quote in question has been taken from a wider context and only gained its anti-ecological character at the end of the Middle Ages—and not as a religious doctrine, but as a philosophical idea. Later, when the paradigm of a scientific-technical civilization was established, adopting the concept of human superiority over nature, it was reshaped into a principle of human domination and unrestricted exploitation of the environment.

Still, Christian ethics do not remain indifferent to ecological issues. Nature is a gift from God the Creator, given to man as an individual, but not without restrictions. Not only are we subject to biological laws towards nature, but also to moral ones. Such a position means identifying the ecological order with the moral one. By violating it, humans break the most important commandment: love for the Creator and also for other people. Moreover, the environment allows humans to fully develop, and thus becomes their partner in a way. So, nature is very strongly connected to humans, thus participating in their dignity (personalism).

How to prevent environmental threats in the light of the above?

Jan Grzesica notes that these threats are usually seen in technical, medical or economical categories, whereas they should be seen in moral categories as well. As he writes: "the point is to stimulate a specific kind of responsibility in humans, responsibility for the world, in which they are living, for their existence and the existence of others. After all, humans are the only beings capable of taking such responsibility" (Grzesica, 2005). So actions leading to environmental protection must be regarded as good, whereas actions leading to its degradation—as bad (Grzesica, 1983). The postulate by this author is complemented with the notion of ecological sin: "when speaking of an ecological sin, we must speak of a human sin and the shaping of proper attitudes must begin with education and sensitization of conscience towards the value of nature" (Grzesica, 2005; Bajda 1999). Ecological sin relates to the situation, when the common welfare is regarded as *res nullius*—nobody's property. It is also one of the components of a new area of science: theo-ecology, which is the theological study of the supernatural God's environment, within which humans live. Natural environment is perceived here as God's sign, a sort of a sacrament (Rogowski, 1992).

Tadeusz Slipko (1985) comments the above conclusions in a slightly different way:

1. The subject of Christian environmental ethics is the human being as an individual.
2. The justified recognition of humans' superiority over nature does not mean absolute dominance.
3. Nature participates in humans' dignity and is strongly connected to them. It must not be regarded as merely a means to the ultimate goal.

These principles should be put in an eschatological context—by reference to the Last Judgement. If humans do not act responsibly, they will be justly punished.

The Christian vision of nature is reflected in numerous documents: from Pope Pius XI's encyclical "Divini redemptoris" of 1937, through the symbolic proclamation of St. Francis, patron of ecologists and Pope John Paul II's encyclical, to Pope Benedict XVI's statements. The main theses of these teachings have been assembled in Table 19.

Table 19. Ecological issues in the light of the documents of the Roman Catholic Church (Zieba, 1995).

No.	Principle
1.	Environment is the fundamental condition for all life, which also applies to humans.
2.	Natural environment is a factor conditioning personality development, as well as cultural and social development of a human being.
3.	Humans must respect the laws of nature.
4.	Earth with all its resources belong to all people of both the present and future generations.
5.	Humans take responsibility for their actions towards the environment both before God and before other people.
6.	It is the duty of every generation to leave the world in such a state, as to provide secure conditions for the future generations to develop properly.
7.	The biblical order to rule the Earth does not entitle us to exploit it relentlessly.
8.	By opposing the Creator's plans, humans brought disorder into nature.
9.	Hazards (including e.g. nuclear potential and chemical contamination) result from an imbalance in nature brought on by the disorderly exploitation of natural resources.
10.	Countering these hazards should include eradicating poverty (regarded as the greatest threat in the environmental context as well), changing lifestyle and a peaceful solution of conflicts between nations.
11.	Environmental hazards cannot be solved only by means of technology. A moral transformation is necessary, developing respect for all life and acknowledging that the world is a common good.
12.	Every religion is responsible for environmental hazards and is assigned a special role on the way to their elimination (within a morally consistent vision of the world).
13.	Environmental degradation means introducing disorder in nature and is inconsistent with the Creator's plan.
14.	Responsibility for nature start at the family level and further includes national and universal level.
15.	Particularly, responsibility for introducing order in nature is assigned to Christians.

Out of the various documents, we should emphasise the encyclicals of John Paul II, who consistently preached about the need to build a civilization of life, peace and love—one, whose values are most consistent with the assumptions of sustainable development.

The encyclical "Redemptor hominis" of 1979 was mostly devoted to widely understood ecological issues, considered in the context of various challenges for modern civilization. It included the following: "The development of technology and the development of contemporary civilization, which is marked by the ascendancy of technology, demand a proportional development of morals and ethics. For the present, this last development seems unfortunately to be always left behind. (…) The essential meaning of this (…) 'dominion' of man over the visible world, which the Creator himself gave man for his task, consists of the priority of ethics over technology, in the primacy of the person over things, and in the superiority of spirit over matter. (…) Man cannot (…) become the slave of things, the slave of economic systems, the slave of production, the slave of his own products."

The issue of determining man's place in the modern world was also raised in "Sollicitudo rei socialis" of 1987 (which is an explication of the humanistic assumptions in the Pope Paul VI's encyclical "Populorum progresio"). It introduced the notion of 'overdevelopment', which regards rich countries and of 'underdevelopment' of the poorer majority. These issues were placed in the environmental context. John Paul II wrote: "one cannot use with impunity the different categories of beings, whether living or inanimate—animals, plants, the natural elements—simply as one wishes, according to one's own economic needs. (…) Using them as if they were inexhaustible, with absolute dominion, seriously endangers their availability not only for the present generation but above all for generations to come. (…) Once again it is evident that development, the planning which governs it, and the way in which resources are used must include respect for moral demands." With it, one must not forget about humans. In his encyclical "Centesimus annus" of 1991, the Pope notes that in addition to the "destruction of the natural environment, we must also mention the more serious destruction of the human environment."

In this context we should also mention the "Message for the World Day of Peace" in 1989, which is the presentation of the Christian ecology. In it, John Paul II wrote: "When man turns his back on the Creator's plan, he provokes a disorder which has inevitable repercussions on the rest of the created order. (…) People are asking anxiously if it is still possible to remedy the damage which has been done. Clearly, an adequate solution cannot be found merely in better management or a more rational use of the Earth's resources, as important as these may be. Rather, we must go to the source of the problem and face in its entirety the profound moral crisis of which the destruction of the environment is only one troubling aspect" (John Paul II, 1989). The Pope also postulates the necessity for the humans to adopt an ecologically responsible attitude towards themselves, towards others and towards the environment. It is worth mentioning that the works of the World Council of Churches, which inaugurated the program "Justice, Peace and the Integrity of Creation", was maintained in a similar vein (Czaczkowska, 1995).

In the case of Pope Benedict XVI, it is worth mentioning his statement delivered in the Vatican on 6th January 2008 during the Epiphany Mass, which placed the discussed issues in the context of globalization. This was predecessor of the encyclical "Caritas in veritate", published in 2009.

The examples above show how the Church perceives environmental issues in the wider context of human moral development. Our trust in technical capacity is not enough to stop the degradation of the world of nature—it also requires a transformation in mankind's spiritual dimension. Such an approach is consistent with the postulates of sustainable development.

However, irrespective of ethical motivation (be it religious or secular, anthropocentric or non-anthropocentric), adopted in the discussion on sustainable development—what matters is the result of the actions taken. What needs stressing, a number of decisions regarding the choices we make depend on each and every one of us individually. This is the practical side of the problem.

3 ETHICS IN PRACTICE: ECOLOGICAL ATTITUDE AND EDUCATION

The moment of implementing theoretical recommendations to everyday activities is a specific sort of 'practical ethics'—regarding individual systems of values, which we respect in our lives. It is a crucial matter, since it is important whether people take a particular action because they are convinced it is the right thing to do, or simply because that is the law.

The major part in the shaping of 'practical ethics' is played by proper education, including the environmental (ecological) education, or education for sustainable development.

According to the UN definition, education is the spread and acquisition of knowledge, skills and values, which allow us to understand the world we live in (UN, 2005). Environmental education may be also described as activities supporting a balance between social and economical welfare and culture, tradition and preservation of Earth's natural resources (Ferrer-i-Carbonell & Gowdy, 2007; Rugumayo, 1987; Smyth, 1987; Greig et al., 1987). This view on education has been supported by the UN General Assembly, which in 2005 accepted the document "Decade of Education for Sustainable Development".

How to conduct a proper ecological education? First of all, one must determine to whom and at what level should it be directed.

Zbigniew Hull distinguishes three levels of ecological education (Hull 1994, 1995):

- Civil level: basic ecological knowledge, necessary for everyone in their everyday life.
- Decision-making level: includes thorough knowledge of ecological contents necessary to make decisions regarding economic, political or other similar activity.
- Expert level: related to the importance of having detailed and continuously updated knowledge necessary for the proper formulation of assessments and recommendations.

Going through each of these stages is a complex task. In case of the basic, civil level, the problem lies in the lack of social interest in the issue of sustainable development. The Decision-making level is associated with politics, where the timescale in question very rarely exceeds the date of the next (local, national, or Union) election. In such conditions even the most solid expert postulates may have problems with crossing over to the other two levels.

A hint in this context may be the philosophical approach, since it considers the issue of ecological education (or putting it more widely: culture) in a different way, namely at the following levels:

- cognitive,
- emotional and volitional,
- behavioural.

Wlodzimierz Tyburski suggests: the first level is related to the spread of environmental knowledge, the second one focuses on inspiring and shaping moral sensitivity to the problems of the world and of nature. The third level shapes actual attitudes and behaviours (Tyburski, 2002).

To be explicit, ecological education should not be limited to the aspect of knowledge. Most likely, most people do realize what is important for the environment. However, good things are not always profitable. Furthermore, nature can still be destroyed, even when everyone has the appropriate knowledge (at the postulated civil level, at least). This situation not only calls for knowledge, but also the people's conviction and engagement for the environment. In short, the point is to inspire an 'ecological consciousness', so an 'ecological attitude' (Aleksandrowicz, 1978).

It should be noted that human attitudes always function in relation to something, and it can be defined as the relatively stable, emotionally reinforced readiness to react to particular persons, groups or situations, in a coherent or consequent way (Zimbardo & Ruch, 1996). In this context, as Wieslaw Sztumski points out, an ecological attitude can be described as readiness to take actions aimed at preserving the more and more endangered and degraded environment, expressed by adopting the proper human-environment relation (Sztumski, 1999).

Such an attitude is worth putting into a wider cultural context (which makes it an ecological-cultural attitude). Danuta Cichy notes that environmental (ecological) culture can be understood as a specific behaviour, based on a system of knowledge, beliefs and accepted values consistent with the principles of respect for all life and for nature (Cichy, 2002) and therefore, through reference to values, it is strictly associated with environmental ethics.

How can such an attitude be inspired? How to conduct an ecological education supporting it?

A number of postulates may be presented in the discussion (Pawłowski, 2005):

- Ecological education is a process, which obviously has a beginning, but it is hard to mark its end. Particular educational processes may be ended, but education itself goes on. Pro-ecological attitudes of children in primary schools do not necessarily correspond to identifying with them in the adult life (Pawłowski, 1996). In order to affect human behaviour, one must simultaneously inspire the attitudes of whole communities, children, youth and adults alike.
- Although information has an important role in education, it is personal experience that is the most significant. It is worth quoting the words of Krystyna Ablewicz: "in order to learn, one must walk a specific path, one must take a trip in order to reach their destination. (…) Experience, of which we learn in this way, is not some ordinary information. Experience requires our presence, information does not. (…) One can inform others of their experience, but cannot transfer it to others" (Biderman, 2002). If experience is so important in shaping attitudes, then projects, which refer to enriching experience and require going outside school walls, gain special importance in ecological education. This makes it easier both to better know nature and to transfer other educational values.
- Ethical postulates should play an important role in ecological education; therefore, all pupils should be subject to compulsory courses on ethics.
- There is no *the* program of environmental education. The point is not only competition (which program is better, and which worse), but also the social group to which it is targeted (variables of age, education, residence). So far, however, the majority of educational programs in many countries are only being prepared for children and youth.
- Ecological education is related to both formal and informal education. It is worth distinguishing new, alternative educational methods. Already in 1972, Jozef Kozielecki suggested that the traditional education system, based on classes and lessons, does not serve human development, openness to dialogue nor control over emotions (Krakowiak, 1997; Baes et al., 1987).

- We should also avoid extremely radical proposals, often called eco-terrorism, or fundamentalist ecology (Slipko & Zwolinski, 1999). Education must not relate to promoting violence.
- Environmental education does not need to be a program that is created completely afresh. Reference should be made to what has already been achieved and proven in previous educational programs. Many ideas that were once thought of as alternative are now part of the educational standard.
- It is particularly worth developing the interdisciplinary path as well as individual subject paths. This would be a duplex path: subject knowledge and environmental knowledge related to it (Cichy, 2002).
- Modern media, especially the Internet and public TV, can help in such education. In practice, however, things look different.
- The issue of the teachers' contact with students is important. It is no mystery that school trips more and more frequently have a purely de-stressing character, where education (ecological and cultural) often recedes into the background.

  In my opinion, the issue of mutual contact depends not so much on the teacher's age, as on their predisposition. However, young people sometimes hardly communicate among themselves, so one does not need to refer to the eternal 'generation conflict'
- The issue of updating knowledge also depends on the teacher. The data is subject to continuous change—proper ecological education must always be up to date.
- Among educational programs such projects as the widely known 'Cleaning up the World', are of great significance. Through participation, young people learn that litter is quick to throw away, but much slower to clean up.
- Didactic tourist walking paths are a very good tool. These usually have a purely ecological character, which involves i.e. forest paths, botanic paths, dendrological paths, fauna paths, geographical paths and general paths, concerning environmental protection (Miskwa, 2005; Borkowski, 2002). However, in addition every region has its historical and cultural context. Nature is threatened, but so are local culture and customs. Just as sustainable development is a multidimensional phenomenon, so is environmental education and didactic paths should be multidimensional as well (so also historical, cultural, or archaeological).
- Official didactic walking paths are important, but much more can be achieved by suggesting that the young people prepare such a path by themselves, hence referring to the local problems and resources. One can use the generally available paths or forest trails, prepare the description using their computers and copy it with a photocopier.

  This is an attempt at a common answer to the question: What can we do in our town, in our park? Stressing these issues prompts intrinsic engagement and identification of the participants. Certainly, some fragments of the path will be already known to the youth, but did they stop by a rare plant's position, or by an old chapel? Yet, knowing one's area, one's 'small homeland', is a very important step towards shaping the attitude of respect—both for nature and for the works of culture.
- Ecotourism (or sustainable tourism) is also important (Honey, 2006); it not only leads to places rich in nature but also in culture. Some significant aspects

are associated with it, such as: the necessity to provide non-invasiveness to the natural environment tourism, rich in elements of ecological education, as well as stressing the benefits that a local community can hold on account of such activity (see "Enviro" special issue 17/1994, entirely devoted to the issue of eco-tourism and the environment).

- Agro-tourism is a specific form of environmental education, related to eco-tourism. The topic is worth touching upon, since food overproduction is a major problem in the EU. In this situation, moving farmers from the production sector to services—such as agro-tourism—gains particular significance.

Furthermore, the farmers who take up such activity often find interest in their area: both in natural and in cultural aspects (in terms of renewing traditions, which previously seemed condemned to oblivion).

Agro-tourism also prompts interest in the 'ecological agriculture'. Therefore, although the decision to establish an agritouristic farm is usually economically motivated (profit-driven), it often contributes to the environmental education of the farmers, and through them—also to the tourists staying at their farms.

Many of the postulates presented above refer to school education. I would also like to point to the enormous role of family in shaping human attitudes towards the environment. This is the most private sphere of human existence. It is also the sphere of the adopted system of values, frequently related to patterns inspired by the parents' personal example. The family should provide social security, of course, but there is more to it—enough to mention emotional security.

In general, it is the family that mostly decides on our quality of our life. It is worth presenting this last notion in more detail, as it refers not only to the context of ecological education, but to eco-philosophy as well. The discussion on the quality of life has a long history; the very notion of 'quality' was first introduced by Plato. He described it as an judgement made by a person as well as some degree of perfection (Rogala, 2003). How about the quality of life?

An important definition was presented by Tadeusz Borys, who notes that "it is a category which integrates all other qualities, which thus become sub-qualities, explaining the nature of life and the bases of its assessment" (Borys, 2003).

Valuation seems to be the key matter here, frequently conducted by summarizing opposites. It can be said that life is good or bad, joyous or sad, happy or unhappy.

Much can be said about the quality of life in the descriptive context, by pointing at the characteristics, which it should have e.g. in the psychological, spiritual, physical, material, biological, or ecological aspect.

These issues interpenetrate even in such fundamental matters as the choice between 'being' and 'having'. It is true that even when focused on 'being', human life requires meeting several basic material conditions, resulting from the biological aspect of our functioning. This is why so many people trying to solve the issue of 'having' or 'being', would answer with the word 'rather'; that is, to rather 'have', or rather 'be' (Tyburski, 2003).

Much depends on one's personal experience and on the social environment we live in. Our individual decisions may affect other people (Stokes, 1981). In other words, action must be taken starting with oneself! Table 20. shows a

Table 20.   How can an individual contribute to environment protection? (Author's own work).

No.	Proposal
1.	Better house insulation. Results in lower energy consumption for home heating, and yields financial benefits.
2.	Replace light bulbs. Standard bulbs are cheap, but consume a lot of energy, thus making their use very costly. Energy-saving fluorescent lamps are an alternative. They are more expensive, but future bills will decrease quickly.
3.	Turn off electronic equipment, when not in use. At present, most equipment does not have the option to shut them down completely; instead, they go into the standby mode, which means continuous energy consumption. It is worth buying a surge protector, to which electronic equipment can be connected and which can be turned off at any time.
4.	Reduce the use of chemicals in your household. Does your white shirt really need to be this white?
5.	Buy products that have been rationally packed. In case of some products, a metal or thick cardboard package serves no purpose and only wastes resources.
6.	Segregate waste or, if this is not possible, demand its implementation. Household waste divides into a dry fraction (e.g. paper, plastics) and an organic fraction (food leftovers). The first one can still be reused, provided it is not mixed with food leftovers. Hence the importance of communal waste segregation.
7.	Carefully tighten the taps and check their tightness. You will use less water and pay less.
8.	Choose drinks in glass bottles instead of in metal cans. They taste better and the bottle can be reused.
9.	Buy local fruits and vegetables rather than imports from the other end of the world. You will eat less chemicals and the problem of transportation over long distances will be eliminated.
10.	When possible, leave your car in your garage (or parking) and use mass communication or a bike/walk.
11.	If you have children, teach them the ecological lifestyle and be an example yourself.

few examples of what everyone can do for the environment and to improve their own quality of life.

4   A NEW RESEARCH FIELD: A PHILOSOPHICAL AUDIT

Previously in this chapter, I have presented the basics of the eco-philosophical side of the issue of sustainable development. Our next task is to relate philosophy more strictly to other dimensions of the development. The new scientific disciplines also go in this direction. Ecology, for instance, is not a purely environmental discipline any more. First, it has been enhanced significantly, giving rise to human ecology, which now has a well-established position and methodology. It can be defined as a science about changes related to human organism, of the structure of population and of organization of the society under the influence of environmental conditions it lives in. Furthermore, not only biological, but also social and cultural characteristics of the human species are included here. The main sections of such an understanding of ecology are (Wolanski, 1992):

- Theoretical basics of the human-environment relation (e.g. the vision of nature, also beliefs and attitudes).
- Biological-social issues related to the human environment (e.g. the flow of matter, energy and information within ecosystems).
- Environmental issues of human biology (e.g. adaptability of organisms and their sensitivity to environmental conditions).
- Cultural adaptive behaviour (e.g. the issues of education).

Such an understanding of human ecology is an extension to the—narrower in meaning—cultural ecology, which studies ways of adaptation of human culture to the environment and relations with other human groups (Campbell, 1995). It also corresponds to numerous other scientific disciplines. By studying human behaviour, it shows association with psychology and eco-philosophy, and out of other problem groups it makes use of the achievements of e.g. geography, economy, anthropology and physiology. It is therefore an interdisciplinary area of science, which may also turn out very useful in the discussion on sustainable development.

However, not only the ecological discussion is currently being widened. Geography too was given a new dimension in the form of humanistic geography. It derives from the reflection over the real universe, including man's place and role in it, which is deep-rooted in the European school of thought (Jedrzejczyk, 2001). This includes the relations between humans and nature, between social and economic activity, between an individual and the society, but also between man and space. Therefore, man himself is the point of reference (both individually and collectively). What is important, this concept of geography gives priority to values over the objective content of scientific geography. In other words, a division is made—which does not occur in scientific geography—into facts and values (also into subjects and objects). The goal is a better understanding of humans as well as the conditions they live in. Since the key part of these conditions is related to the environment, humanistic geography also takes the interpretation of natural phenomena. In terms of methodology, it uses methods known in sociology and in philosophy. This is because it is impossible to attempt to determine man's place in nature without including the philosophical context.

The presented examples of disciplines combining ecological issues with geographical or philosophical issues prove that the interdisciplinary approach is more widespread than it seems.

Interdisciplinarity is also included—which has already been mentioned in this work—in the concept of sustainable development, with philosophy as the base. This gives us an opportunity to penetrate completely new research fields.

At present, there are attempts to 'measure up' sustainable development with the use of various systems of indicators. Unfortunately, this lacks something important. According to the domain criterion, the audits mentioned only refer to three scopes of the problem: environmental, social and economic. For instance, the first case may involve asking the following questions:

- What activity does the evaluated unit conduct?
- Can the effectiveness of the conducted activity (e.g. in the aspect of resource consumption) be increased?
- What is this particular plant's impact on the environment?
- What goals does the given unit realize towards environmental protection?
- What are the effects of these actions?
- How is environmental management realized?

The list should be expanded, however, and I would call this expanded version of the diagnostic tool in question—a philosophical audit.

I do realize that philosophical issues are not so easy to measure as e.g. particular pollution indicators in the environment. Nevertheless I think it is advisable to ask questions regarding e.g. the issue of values, when assessing any particular action strategy:

- What values does this strategy include?
- Are they 'compatible' with the concept of sustainable development?
- If so, does the strategy promote these values?
- How?
- Are these values actually realized?

The last question is worth placing in a wider context. It is assumed that one of the ways of shaping attitudes 'compatible' with the principle of sustainable development is the proper ecological education. Although educational programs are evaluated regularly (whether officially or by independent actors), the assessment remains incomplete. In terms of the proposed philosophical audit, the assessment should also include the question to what extent these programs affect the people subjected to education processes. Does their attitude towards the environment change? If yes, for how long? Why would some people eventually change their approach?

A questionnaire method is undoubtedly one of the tools that would come in handy when answering these questions.

Furthermore, many more diagnostic methods may—and should—be used in the philosophical audit. Every scientific discipline, within which the problems of sustainable development are being discussed, may introduce its own methods and propose an original approach that allows including philosophical issues in the discussion. That is why I hope this new and interesting research field would eventually meet with scientists' interest and be taken up more energetically.

After philosophy's division from science, the time has come for its return. Philosophy forces us to make the fundamental reflection, due to which we are able to do more than just implement yet another environmental protection program. I am positive it is philosophy that would enable the full integration of other problem groups of sustainable development, which, in practice, determines the very feasibility of implementing this development.

CHAPTER 4

Level II of sustainable development: Ecological, social, and economic considerations

When speaking of the strategies adopted within the UN and EU, three primary pillars of sustainable development are usually mentioned—the ecological pillar, the economic pillar, and the social pillar. This part of our work will focus on reviewing the essential issues of each of these problem groups.

1 ECOLOGICAL PLANE

1.1 *Natural environment*

The natural environment functions through the exchange of matter, energy, and information between its individual parts, called geocomponents. The most important among them are air, water, soil, plant life, animals, and humans (Richling & Solon, 1994). Protecting this potential is a major, though not the only, task for the ecological plane of sustainable development. Proper land management of the areas already transformed is also essential, both in rural and urban environments.

The discussion should be opened with the theme of environmental protection (Weigle, 1996). This notion holds the ideas of preserving the balanced use and renewal of the resources, formations, and components of nature.

A wider view on the matter was proposed by Adam Wodziczko in 1948, highlighting the three major trends in environmental protection (Wodziczko, 1948):

- The conserver approach. This approach was dominant in the interwar period and related to the protection of surviving precious natural habitats.
- The biocentric approach. This approach puts the discussion in the context of entire biocoenosis systems. The necessity to maintain a balance in nature was also highlighted, hence the need to consider nature's requirements in economic activities.
- The planning approach. Such an approach stresses the importance of systems larger than biocoenosis—the landscape level is treated as an integral biological entity.

Wodziczko also sees the basic goals for nature conservation in a threefold manner. The goals are (Wodziczko, 1948):

- Conserver goals: preserving the remaining enclaves of wildlife. The motivation for protection includes idealistic issues, scientific issues, aesthetic issues, and historical-memorial issues.

- Socio-economic goals: maintaining the 'life force' of nature, which is particularly important in the case of the economic use of the environment.
- Landscape goals: referring to planning tasks and related to the restoration of devastated landscapes, also in human places of residence.

It needs to be pointed out that nature conservation is a specific race against time. A devastated environment can never be restored to its former state; the same applies to genetic resources of extinct species, which are irretrievably lost.

The level of our pressure on the environment is expressed by a number of indicators, of which the Living Planet Index (LPI) is worth mentioning. It shows how, within a single generation (since 1970), 30% of the world's natural resources were destroyed (Living Planet Report, 2006). Despite this quite apocalyptic image, many enclaves of natural types of environment still remain, which makes it all the more necessary to protect them. This is achieved through establishing national parks and nature reserves and by introducing other forms of nature conservation.

The basic ecological unit that can be put under protection is the ecosystem. This fundamental notion can be defined (Krebs, 1998) as all the living organisms (biocoenosis) living in a particular space and time with their inanimate surroundings (biotope). A typology of forms of ecosystems is presented in Table 21.

The placement of an ecological unit in the hierarchy of a biosphere is as follows (Hluszczyk & Stankiewicz, 1996):

- protoplasm,
- cell,
- tissue,
- organ,
- system of organs,
- organism,
- population,
- biocoenosis,
- ecosystem,
- landscape,
- biosphere.

In practice, the most visible result of protective actions is the formulation of environmental protection laws.

On a global scale, these actions are based on three documents (discussed in more detail in Chapter I of this work). These are:

Table 21.   Typology of ecosystems (Author's own work).

Ecosystems	Aquatic	Terrestrial
Natural	Seas and oceans	Forests
	Rivers	Grasslands
	Lakes	Deserts and semi-deserts
Artificial	Ponds and pools	Plantation forests
	Channels (e.g. irrigation channels)	Pastures
	Dam reservoirs	Fields
		Orchards and gardens

- The IUCN's "World Conservation Strategy" of 1980.
- "The World Charter for Nature" of 1982.
- "The Convention on Biological Diversity" of 1992.

These documents contain general recommendations. A more detailed characterization is the classification of protected areas introduced by the International Union for Conservation of Nature (IUCN)—see Table 22.

Table 22.   Categories of protected areas according to the IUCN (UICN, 2010).

Category	Description
Ia	Strict Nature Reserves.
	Strictly protected areas set aside to protect biodiversity and also possibly geological/geomorphical features, where human visitation, use and impacts are strictly controlled and limited to ensure protection of the conservation values. Such protected areas can serve as indispensable reference areas for scientific research and monitoring.
Ib	Wilderness Area.
	Protected areas that are usually large, unmodified or slightly modified, retaining their natural character and influence within permanent or significant human habitation, which are protected and managed so as to preserve their natural condition.
II	National Park.
	Protected areas that are large natural or near natural areas set aside to protect large-scale ecological processes, along with the complement of species and ecosystems characteristic of the area, which also provide a foundation for environmentally and culturally compatible, spiritual, scientific, educational, recreational, and visitor opportunities.
III	Natural Monument or Feature.
	Protected areas that are set aside to protect a specific natural monument, which can be a landform, sea mount, submarine cavern, geological feature such as a cave or even a living feature such as an ancient grove. They are generally quite small protected areas and often have high visitor value.
IV	Habitat/Species Management Area.
	Protected areas that aim to protect particular species or habitats and management reflects this priority. Many such areas will need regular, active interventions to address the requirements of particular species or to maintain habitats, but this is not a requirement of the category.
V	Protected Landscape/Seascape.
	A protected area where the interaction of people and nature over time has produced an area of distinct character with significant, ecological, biological, cultural and scenic value: and where safeguarding the integrity of this interaction is vital to protecting and sustaining the area and its associated nature conservation and other values.
VI	Protected area with sustainable use of natural resources.
	Protected areas that conserve ecosystems and habitats together with associated cultural values and traditional natural resource management systems. They are generally large, with most of the area in a natural condition, where a proportion is under sustainable natural resource management and where low-level non-industrial use of natural resources compatible with nature conservation is seen as one of the main aims of the area.

Most of the protected areas were created in North America (as much as 27.08% of the land area of the United States), while the world's largest single protected area is the Greenland National Park, which covers 972,000 km^2. In Europe, the largest area is the Russian Yugyd Va National Park, covering the area of 18,917 km^2 (Walczak et al., 2001).

In general, the number of all protected areas in 2008 was over 120,000 covering an area of 21 million km^2, which is 12.2% of all land area and 6.4% of the marine and oceanic areas (UNEP, 2010).

The issue of nature conservation is also raised by internal European regulations. The most important of these include the following two:

- "The Maastricht Declaration on Conserving Europe's Natural Heritage" of 1993 (Kozlowski, 2005). It points out that one of the necessary conditions for achieving sustainable development is to stop the devastation of nature.
- "The Pan-European Biological and Landscape Diversity Strategy". Its introduction was discussed during a summit of the European Ministers of the Environment in Sofia in 1995, and a resolution was passed during the next summit in Aarhus, Denmark in 1998 (Environment for Europe, 2010).

This latter was a detailed action plan, which considered the following pillars:

- Level of naturalness (reference to areas free from human impact or ones where such impact is insignificant).
- Diversity (both in terms of species and the habitats they occupy).
- Representativeness (for the particular region).
- Level of threat for species occupying the particular region.
- Size of the area, which is to be put under protection (large objects are preferred, as they make it easier to maintain ecological balance).
- The issue of endemism (paying attention to species living only in one region, or totally absent outside Europe).
- Biological/cultural diversity (combining natural issues with the cultural identity of the region's residents).

The above documents postulate the need to integrate nature conservation with other forms of human activity, particularly the economy. The importance of databases was also stressed. Without them, it would be impossible to gather up-to-date information on the environment. Among them, the ones most worth pointing out are:

- CORINE databases,
- ECONET network,
- NATURA 2000 network.

The CORINE (Coordination of Information on the Environment) program was introduced in Europe in the 1980s. The governing body was the European Environmental Agency (EEA). Three main departments of the program are (EEA, 2010):

- CORINAIR (originally CORINEAIR); related to the protection of the atmosphere, which is perhaps the most important problem group of modern environmental protection (Sully & Hill, 2006).

- CORINE biotopes; a database of valuable and distinctive ecological units, discussed regionally, but also at the European level. Such units may be an entire mesoregion (e.g. a mountain range), or a single object (e.g. a cave).
- CORINE land cover; a program that classifies individual forms of land management and shows the transformations they undergo over time. It is an important tool in the proper shaping of spatial order.

The CORINE databases were used in the creation of the ecological networks of ECONET, and then of NATURA 2000.

The ECONET network was established under the IUCN and was ratified by the Council of Europe in 1992. Its major goal was to identify the most valuable natural areas in individual countries (even now protected with the use of various forms of nature conservation, or planned to be taken under protection) and link them with ecological corridors, facilitating species migration and providing better protection for their habitats (Liro, 1995). The selection criteria were consistent with the previously discussed "Pan-European Biological and Landscape Diversity Strategy".

The basic functional elements of the ECONET network are as follows (Liro, 1995):

- Core areas: consisting of biocenters (where the greatest diversity of natural forms is found) and buffer zones (which serve as support for the protective functions). It has been assumed they each such centre should cover more than 500 ha.
- Ecological corridors: preserving the natural migration of species. These do not have to be uninterrupted, e.g. in the case of birds, the only thing that matters is their functional continuity (Fortman, 1984).
- Nature development areas: areas which have previously been degraded, and whose natural value can be restored (e.g. forests, or agrocenosis).

It should be pointed out, that the ECONET network was a radical project which introduced significant changes to previous principles of action for the conservation of nature. However, its full implementation was abandoned. One of the reasons was that the postulated protected areas represented a significant proportion of Europe—covering 50% of its territory.

Another network—one which was actually created—was NATURA 2000. It restricted the discussion to particular refuges and habitats, skipping the issue of ecological corridors. Its territorial range is consequently smaller, oscillating around 10 to 15% of Europe (Natura 2000, 2011).

The governing body of NATURA 2000 is the European Commission, which oversees the creation of the network, based on the following directives:

- Directive no. 409 of 1979 on the protection of wild birds (the 'Birds Directive').
- Directive no. 43 of 1992 on the protection of natural and semi-natural habitats, and wildlife (the 'Habitats Directive').
- Directive no. 266 of 1996 on the scope of information on areas proposed for the NATURA 2000 system.

The practical aspect of the network consists of the creation of two types of protected areas:

- SPA—Special Protected Areas, established based on the 'Birds Directive'.
- SAC—Special Areas of Conservation, established based on the 'Habitats Directive'.

Apart from that, a proposal has been prepared, to establish areas which, upon acceptance by the European Commission, would be given the status of refuges of Sites of Community Importance (SCI).

The NATURA 2000 network was supposed to create a new integrated system of environmental protection in Europe. This would have happened, had the previous, more radical European project, the ECONET network, been finished. The NATURA 2000 project, adopted in its place, was a compromise solution and, as such, may turn out to be just another version of area nature conservation. Its major weakness is that it is almost completely detached from previously introduced traditional forms of nature conservation. The creation of traditional forms has a long history (as presented in Chapter 1) and, although there are significant differences regarding the names or levels of protection, every European country has its own well-established protection methodology, proven over years of practice. Of course, the NATURA 2000 network does not eliminate previous forms of protection. Furthermore, the newly established areas may overlap with these. However, the postulated uniform management seems unlikely, since it is impossible to unify the methods of protection for objects that, according to law, are national parks and nature reserves, which are areas of high-level protection, with such objects as landscape parks, where restrictions are much less severe. Conflicts are therefore inevitable. Some actions that are forbidden in national parks, for example, are allowed within landscape parks; and creating a NATURA 2000 object within the same area lays out a 'third path' of action. The very names of the areas can be misleading as well, as they are almost identical—Special Protection Areas, Special Protected Areas.

Figure 2.   Poland: The largest natural deciduous lowland forest in Europe, Bialowieza National Park. Photo by Artur Pawłowski.

Figure 3.   Austria: Grossglockner Glacier, High Tauern National Park. Photo by Artur Pawłowski.

Figure 4.   Italy: Tofana di Rozes in Dolomiti d'Ampezzo National Park. Photo by Artur Pawłowski.

But despite its deficiencies, the major merit of the NATURA 2000 network is that it shows Europe's vast natural diversity and its main hazards. It is the first attempt at unifying systems for protecting valuable areas throughout the continent; this, in the long run, may turn out to be a breakthrough in the history of environmental protection. The pictures illustrate a selection of several environmentally valuable protected areas in Europe.

Figure 5.   Poland: Sand dunes in Slowinski National Park. Photo by Artur Pawłowski.

Figure 6.   Slovakia: Sucha Bela gully in Slovakian Paradise National Park. Photo by Artur Pawłowski.

1.2   *Changed landscape*

The discussion concerning the ecological plane of sustainable development goes beyond nature conservation and refers to managing space as well, particularly that already transformed by humans at the landscape level. According to

V. Vanick, it is a continuous process leading to the optimization of the ways in which to use particular areas of the Earth, while preserving their productivity and beauty (Richling & Solon, 1994).

In line with this description, the landscape itself would be the 'particular area of the Earth'. Putting it more clearly, the landscape is part of the epigeo-sphere, i.e. the outer part of the Earth, which is a complex system consisting of forms, reliefs, waters, plant life, soils, rocks and atmosphere.

Its features are (Richling & Solon, 1994):

- It occupies a particular section of space, which can be presented on the map or photographed.
- It has specific surface features.
- It is a dynamic structure.
- It undergoes evolution.
- It is affected by humans.

The division of landscapes may include their ecological structures or the extent of human impact. For the former, the following landscapes should be distinguished (Richling & Solon, 1994):

- Lowlands.
- Highlands and low mountains.
- Middle and high mountains.
- Valleys and depressions.

In terms of the level of human impact, we can list the following possibilities (Dobrzanski, 1993):

- No interference.
  It is a theoretical level, as practically no area has been left untouched by human hand, if only in terms of air pollution or climate change. However, undoubtedly there are still areas with a high level of naturalness.
- Disturbance.
  This is the level of interference that does not destabilize the self-regulating mechanisms of an ecological system (e.g. ecosystem).
- Damage.
  At this level, the ecological system has been changed, although nature is still able to restore its previous state with the help of natural mechanisms, e.g. ecological succession.
- Devastation.
  Interference is severe enough that nature is not to able to restore its former state, or the process might take many years.
- Change of natural systems to cultural ones.
  It is interference that permanently and intentionally transforms ecological systems. The state of change is being continuously and artificially main-tained. These are cities, or agricultural environments.

The last level of interference is a purposeful activity, which helps in main-taining man's own infrastructure. All the other levels are placed between rel-atively low interference (natural landscape) and a massive degradation of the environment (devastated landscape). However, the intermediate stages are the most common. Moreover, the issues of the ecological structure of the environ-ment and the scope of human interference overlap. These aspects have been included in the proposal for landscape divisions proposed by Frantisek Kele and Peter Mariot—see Table 23.

Can the level of human interference in the landscape be limited? The basic action strategies in terms of landscape shaping are similar to those of environmental protection. Taking into account the increasing scope of possible human interference, they would include various forms (Richling & Solon, 1994):

- Conserver protection of the landscape; allowing no transformations. This is strict nature conservation.
- Revalorization of landscape; which refers to increasing the natural value of a given area. It may include introduction of rare species of plants and animals to new habitats.
- Landscape management; preserving its current value. In this case, introducing changes is allowed and even necessary at times.
- Restoration of a landscape that has been damaged. Here it is necessary to take preventive action, aimed at restoring the destroyed elements of the ecosystem, thus returning it to its natural state.
- Restoration of devastated landscapes. It seeks to restore the basic functional elements of an ecological system which are significant for human economy, such as reconstruction of its biological productivity (e.g. for agriculture).

It needs to be stressed that the proper landscape shaping leads to achieving spatial order, which is the fundamental agent of the ecological plane of sustainable development. This order can be described as "such shaping of space, which makes a harmonious whole and considers in an orderly manner, all the conditions and functional requirements: socio-economic, environmental, cultural, and composition/aesthetic." This interesting definition, coming from the Polish "Act on Planning and Spatial Management", introduced in 2003, outlines a broad context, which, in its essence, includes most of the problem groups regarding sustainable development.

Much help in the discussion on spatial order can be found in the CORINE Land Cover database. It distinguishes five elementary forms of land cover (Ciolkosz & Bielecka, 2005):

Table 23. Divisions of landscapes including the ecological structure and scope of anthropogenic interference (Kele & Mariot, 1986).

Groups of landscapes	Types of landscapes
Natural landscape not used economically	Arctic
	Alpine
	Desert
Natural landscape with potential opportunities for use	Humid rainforests
	Taiga
Extensively used natural landscape	Tundra
	Mountain meadows
Transitional landscape extensively used	Deciduous forests
	Savannah
Rural cultural landscape	Agricultural
	Recreational
Urban cultural landscape	–

- Anthropogenic terrain.
- Agricultural lands.
- Forests and semi-natural ecosystems.
- Wetlands.
- Water areas.

This classification is similar to the work taking up the issue of environmental diversity at the landscape level. The first group is particularly significant here, as it includes the most transformed areas, i.e. industrial and communication areas, urban buildings, urban green areas, mines, and exploitation hollows—see Table 24.

Agricultural lands form a separate group, since agriculture is a very significant aspect of space shaping, which deserves a more detailed inquiry.

In this context it is worth highlighting the positive impact of a new trend, associated with the creation of 'organic' or 'ecological farms' (Woodward et al., 1996; Ikerd, 2008). This form of agriculture is based on the "Regulation of the European Council no. 2092 of 1991 on organic production and labeling of organic products".

A number of principles can be distinguished within this vision of agriculture (Bodin & Ebberstein, 2007):

- Integrated land cultivation and animal husbandry. The manure obtained from animals is used within the very same farm as fertilizer for crops.
- Methods of husbandry consistent with the needs of particular species.
- Non-use of pesticides.
- Non-use of artificial fertilizers.
- Non-use of growth regulators, synthetic preservatives, or artificial food improvers.
- Using only natural components (manure, compost, green fertilizers) and natural methods (such as crop rotation).
- The additional condition for recognizing food as ecological should be whether it has been manufactured locally. The popularity of the slogan 'ecological' has drawn the attention of large companies to this market sector.

Table 24. Anthropogenic forms of land cover found in the CORINE Land Cover database (Ciolkosz & Bielecka, 2005).

Level I	Level II	Level III
Anthropogenic areas.	Urban buildings.	Dense urban buildings. Light urban buildings.
	Industrial areas, commercial areas and communication areas.	Industrial or commercial areas. Communication and areas related to road and rail transport. Ports. Airports.
	Mines, exploitation hollows, and construction sites.	Opencast exploitation hollows. Waste and mine dumps and tips. Construction sites.
	Urban green areas and recreational areas.	Green areas. Sports and recreation sites.

The food manufactured by them, although usually made with a limited use of chemicals, is, nevertheless, being transported over vast distances. Moreover, farmers are forced to specialize, which often means cultivating only one species (monoculture) in as big an area as possible. This violates the fundamental principle of ecological agriculture, relating to the necessity of simultaneous land cultivation and animal husbandry (Zwawa, 2005).

Some of these assumptions are subject to discussion, e.g. plants need fertilizers for growth, but whether the nitrogen and phosphorus come from natural or artificial sources seems not that important. This is because the problem with synthetic fertilizers is not that they are manufactured in factories, but that they are often excessively used in large quantities. Nevertheless, the organic farms are usually set up in regions with a healthy environment and the food produced at these farms is of high quality.

The share of organic farming in Europe in the total utilized agricultural area is (Eurostat, 2009) the highest in Austria (18,5%, the UE average is 5,7%), while the total organic area (fully converted or under conversion) is the highest in Spain (1 317 539 ha) and Italy (1 002 414 ha).

We might want to recall the general belief that ecological agriculture gives smaller yields than traditional one. This is true in many wealthy areas, where agriculture has become heavily technological. However, as studies show, in poor countries, such as Cambodia, the application of the principles of ecological agriculture has contributed to a 71% yield increase (Cooley, 2006).

The opposite of ecological and traditional agriculture is the 'factory farming' carried out by large companies (Sierra Club, 2005; Clunies-Ross & Hildyard, 1996). At first they operated only in the USA, but they have also been building facilities in Europe for the last several years, e.g. in Poland and Romania.

These companies' products are hard to spot as they are sold under various brand names (the companies deliberately buy up a number of businesses in order to gain new brands, usually not associated with factory farming). During an American food control undertaken in Washington in 2003, it turned out that one of the supermarkets theoretically provided articles manufactured by 50 different companies, which however all belonged to a single large consortium (Civil Action, 2007). Moreover, the massive scale of the problem is best shown by the sharp decline in the number of traditional farms observed in the USA. Within the last two decades of the 20th century, their number dropped from 600,000 to just 157,000, at the same time meat production was kept at the same level (Lang, 1998). This means that thousands of farms have gone bankrupt or been taken over and merged with the large breeding farms owned by the consortia.

The essence of the problem lies in the very name 'factory farming', which means massive crop cultivation or breeding thousands of animals crowded into a small area. Such solutions are profitable, as the food produced in such a way is very cheap.

However, the economic aspect of the issue cannot override the health and environmental aspects.

Health aspects are mentioned here, because animal density favors the development of diseases, which forces the addition of antibiotics to the fodder. As a result, strains of bacteria appear, that are immune to antibiotics; this poses a direct threat to the health of humans.

Environmental aspects are mainly related to the enormous amounts of waste. In the case of traditional farming, these are often used for fertilizing crops. In the case of industrial breeding farms, however, there are no crops. Moreover, it is hard to imagine how extensive the crop areas themselves would have to be in order to consume such amounts of waste. For instance, in North Caroline, at one of the many farms belonging to one of the biggest companies, Smithfield, 2500 pigs are bred. Such a herd would generate almost 100 million liters of liquid excrement, 80 million liters of mud, and 4.5 million liters of other waste (Public Citizen, 2004). Furthermore, the so-called 'dung' coming from these farms may contain more than 100 times the pathogens content of human excrement (America's Animal Factories, 1998). It also happens that the containers used for storing the waste do not fulfill their role, resulting in environmental pollution, particularly soil and groundwater pollution (Sierra Club, 2010).

These farms can also lead to large-scale disasters. In the summer of 2007, an epidemic of swine fever broke out on Romanian farms belonging to the same company. As a result, a decision was made to close 11 out of 33 farms. Moreover, the company has been given a penalty of € 130,000 for failure to comply with sanitary standards; cases of veterinary negligence have also been reported (Rolnicy.com, 2007).

Fortunately, there are efficient methods of putting pressure on companies to force them to increase their environmental standards. It seems that much can be done, especially by legal means, which is why we will return to this topic later on, when we discuss the legal factors of sustainable development.

Another example of some significant errors made in the development of agriculture is Bovine Spongiform Encephalopathy (BSE). It was caused by humans breaking the laws of nature. In order to reduce costs, cows were given fodder with added meat-and-bone meal instead of natural plant components. This meal was made from the carcasses of dead animals, including sheep. The companies ignored the fact that many of the sheep had died of an illness called 'scrapie' (whose symptoms include the sick animals constantly scratching themselves), which is caused by the presence of prions, a type of infectious protein. As a consequence, prions also infected the cows. The first case was identified on February 11, 1985, a year later there were 62 cases, and in 1992, 36,000!

It is worth mentioning that at first it was said that the disease was caused by viruses. A breakthrough came with the work of an American, Stanley B. Prusiner, who pointed to prions. Initially his reports were rejected by other researchers, until more and more facts supported it over time. For that discovery, Prusiner was eventually given a Nobel Prize in 1997.

As it turned out, people eating infected beef were also in danger. Although BSE had been diagnosed before, as Creutzfeld-Jakob disease, it had been very rare and the disease was characterized by a long incubation period (sometimes lasting 30 years). Eating infected beef resulted in this incubation period becoming a great deal shorter, and the disease threatened young people as well. It is enough to say that scientists had started to talk of an entirely new variety of the disease. It attacked people in 1996 in Great Britain and killed at least 100 within the first five years. We say 'at least' because the disease is often misdiagnosed (e.g. it is confused with Alzheimer's disease or with mental disorders). It is impossible to estimate the number of people who are infected, but in whom the disease has not yet developed. It is worth mentioning that the infection may not necessarily come from eating beef. There are many products (gelatin, medicines, vaccines, cosmetics) that use substances coming from cows. This is why it is not possible to predict how this threat will develop in the years to come.

The mistake of ignoring the laws of nature by giving animals food they would never eat in natural circumstances was not the only one that was made.

Another bad decision was to ignore the problem, despite the fact it had been recognized. Even in 1990, the U.K. Minister of Agriculture, John Gummer, argued that British beef was perfectly safe; his words were reinforced with the picture of his four-year old daughter eating a beef hamburger. In this context, it is not surprising that the decision to kill herds was late in coming and was carried out inconsistently. What is terrifying, according to the European Commission, is that while some herds were being eliminated, others were being exported to other countries. As many as 60,000 cows travelled outside the borders of England between 1985 and 1990!

This narrowly focused action was an attempt to reduce the losses of the beef manufacturers. As a result, however, the disease spread across Europe and the costs of eliminating the additional herds that became infected were much larger than what would have been paid out if radical action had been taken as soon as the threat had emerged.

Agriculture and some of problems associated with it, which are presented above, are not the only possible ways of shaping the natural environment.

Let us recall a group of anthropogenic terrains that are listed in the CORINE Land Cover database. These are the most burdensome ways of land management, such as industrial plants, or cities. Their impact on the environment goes far beyond their physical territory. It should be highlighted that, although the issue of industry-related environmental degradation is more and more controllable (especially on the technical plane, which will be shown in a further part of this book), the development of cities is often chaotic and constantly running out of control. This part of the work focuses on the factors of the urban environment, with special concern for their ecological dimensions.

The most fundamental European documents taking up the issue of urban development include:

- "The European Regional/Spatial Planning Charter" ('Torremolinos Charter') of 1983, which is a set of general declarations, discussed in more detail in other documents.
- "The Green Charter for Urban Planning" of 1990, which is an official document of the European Commission. It focuses on the issue of the ever-increasing growth of cities and regulating the ecological aspect of building in new areas.
- "The New Charter of Athens—a Vision for Cities in the 21st Century" of 2003, produced by the European Council of Town Planners. Taking up the issue of proper spatial planning (described as a fundamental, yet limited natural resource, demand for which is constantly increasing) a vision was proposed, of the 'cohesive city', whose important element would be environmental cohesion, i.e. harmonizing the urban environment with the natural one. Economic and social cohesion have also been recognized as important, which is in line with the assumptions of sustainable development.
- The informal "Aalborg Charter for Sustainable Cities and Towns" of 2004 refers to intensified construction, transportation, and energy systems, as well as promoting the multifunctional use of particular terrains.

Information provided by new scientific disciplines, such as landscape ecology, is also helpful. Its goal is to identify and prioritize the elements of the landscape, taking into account the relations between them (Truner et al., 2001; Richling & Solon, 1994). Humans are also considered elements of the landscape, and the results of their actions are subject to evaluation in terms of their impact on the organization of natural space. It is therefore an issue of urban environments as well.

The problem of proper urban development was discussed in the 16th century. The year 1580 saw the introduction of a regulation strictly prohibiting the building of houses within three miles of the walls of London (Borhulski, 2006). This, however, was an isolated case document, and the subject was more elaborately discussed at the turn of the 19th and 20th centuries (Dylewski, 2006). The major problems remain the same; the only thing that had changed was their magnitude. At the beginning of the 19th century, only 3% of people in Europe were living in cities. Nowadays, the degree of urbanization has reached 85% in the west and 70% in the east of this continent (Anderson et al., 1997). These changes are significant at the global level as well. In 2008, for the first time in history, the urban population exceeded 50% of the total (World Watch, 2008). Moreover, we have 20 cities with over 10 million residents in each.

This increased degree of urbanization results in a number of ecological consequences, one of which is the issue of waste storage. This problem is getting increasingly worse. For instance, the landfill of municipal solid waste, perfidiously named Fresh Kills, located on Staten Island, New York, covers 12 square kilometers and is 25 metres higher than the famous Statue of Liberty standing nearby. It is one of the largest artificial objects in the world; comparable to the Great Wall of China (Meara Sheehan, 2001). It is striking that one of humanity's greatest creations consists of garbage. The landfill was closed in 2001, however, and a decision was made to reclaim the area and establish a huge city park in that location.

From an ecological perspective, the very structure of buildings is also of importance—their number, their degree of material consumption, and their exploitation parameters. It is worth mentioning the American LEED (Leadership in Energy and Environmental Design) system of certification of environment-friendly buildings, developed since 1994. Its main goal is to reduce the rate of energy consumption, thus decreasing the impact of buildings on climate change. Up to 2007, the program included 14,000 projects within the USA (LEED, 2008). These actions are supported by the American Institute of Architects and the AIA2030 Challenge program (AIA, 2007).

However, the most visible symptom of the impressive changes in the urban environment is different. It is the rapid increase in the area occupied by cities ('urban sprawl').

The turning point in the discussion around the phenomenon is considered to be "Garden Cities of Tomorrow" the work of Ebenezer Howard in 1902. Among the ways to counteract the excessive sprawl of cities, only ecological aspects were being highlighted at first, e.g. the proposed creation of the so-called 'green belts'. These were to be areas around the city that would be left free from buildings (Szulczewska & Cieszewska, 2006). The need to establish green areas within the cities themselves is also stressed (National Charter ..., 1994). In general, we can assume that the main principle of the idea of 'green belts' is to protect biological terrains against urbanization. This principle is realized by undertaking the following tasks (Szulczewska & Cieszewska, 2006):

- Taking control of the excessive and often chaotic growth of built-up terrains.
- Limiting the merger of neighboring urban units.
- Protecting rural areas from excessive building.
- Protecting the spatial conditions of historical cities.
- Restoring degraded urban areas.

Another important challenge is the fact that the best ecological/health conditions are usually found outside, or on the outskirts of cities. Accommodation is usually in the form of single-family homes and the population density is low. Additionally, access to open spaces and green areas is easier. It is, therefore, important to properly join architectonic spaces with the elements of the natural environment, thus enhancing the biological and health levels of the particularly unkempt inner areas of cities (Peski, 1999).

The actions outlined above favor the preservation of the natural landscape outside the cities and makes the landscapes within them more natural (Kostecka et al., 2007). Unfortunately, in many regions of the world, the situation is still far from that. The 'wild' building, observed especially in the suburbs, poses a threat. A particularly serious problem is that these buildings are often built where the green belts around the city were originally planned to be (Radziejewski, 2006). This results in degradation of the natural environment, and in many cases, the agricultural one as well.

What's more, the global issue of the rapid expansion of cities means additional costs—those associated with the necessary extension of the infrastructure. The distances between where the residents work, where they live, and where the offices and stores are, increase. As a result, communication routes also grow longer. This may lead to a failure of public communication and an increased share of private cars in the traffic, which in turn results in traffic jams during rush hours and increased levels of air pollution.

To sum up this part of our discussion, it is worth asking: What does the future have in store for us?

The specifics of the ecological plane of sustainable development are that restoring a once devastated environment turns out to be difficult and expensive and, in some cases, completely impossible. As things currently stand, during the 20th century alone, 22.5% of the land area has been devastated. Of this, one-third is light degradation (Sherr, 2001) and we still lose about 0.6–0.7% of the world's available soil every year (Michna, 1993). In other words, a larger area is being degraded, than that taking part in soil generation. This is a major challenge.

In case of nature conservation, we must hope that a break will be made in the present trend of reducing the level of biodiversity in many regions of the world, including Europe.

In case of human-transformed environments, many issues remain unsolved. Urbanization is one such issue. It seems that the phenomenon would have to be stopped at some point; the whole planet cannot be turned into one enormous city.[3]

---

[3] It is worthwhile looking at the science fiction visions. The famous "Star Wars" saga by George Lucas is the best example. The capital of the Republic, the planet Coruscant, is entirely covered with multi-layer buildings. Without nature, however, human existence is not possible if only for the ecological factors related to the supply of oxygen and food.

This problem goes beyond the ecological plane of sustainable development. Whether people would settle in cities or outside them, is associated with social agents as well.

2 SOCIAL PLANE

2.1 *Social environment*

As with the natural environment, the social environment can also degrade. The environmental conflict is also a type of the social conflict, as it refers to the situation when the conditions facilitating the optimally comprehensive development of a human being become a rarity (Runc, 1988). It is no coincidence that one of the new scientific disciplines, euthyphronics (*euthryphron*—right-minded, sincere) proposed by Jozef Banka, deals with the problem of human psychological immunity to the processes taking place in the world of scientific-technical civilization (Banka, 1996).

The strong connection between the ecological and social aspects is enough to also include the second problem group in the discussion on sustainable development (Harris & Goodwin, 2001).

The notion that humans are undoubtedly social beings should serve as our starting point. Many factors contribute to our environment—customs, culture, spirituality, interpersonal relations, and living conditions (not only economic, but also those referring to the place of residence: in a city or in the country). Even the human-nature relation has its own social aspect, since all human activities toward the environment are always performed through the socio-cultural models in force. Also, environmental issues refer to social science in their formal aspect, but only when they affect humans, e.g. in terms of the hazards associated with them, or the very fact of the humans' noticing these hazards (Leory & Nelissen, 1999).

It is worth pointing out that reflection on environmental issues took place in social science much later than in other sciences, both the natural and technical sciences. Although such terms as 'human ecology' or 'social ecology' were already known in the 1920s (Albinska, 2005), they did not include the full set of relationships between the social system and the natural one. As Riley Dunlap and William R. Catton (Dunlap & Catton, 1992; Lerou & Nelissen, 1999) say, even at the turn of the 1970s and 1980s, prestigious sociological magazines, such as the "American Sociological Review" or "American Journal of Sociology", practically omitted these issues. This may partly result from the rather late formalization of social science. To a large extent, separate academic departments for individual disciplines within this group started to appear after World War II (Wallerstein, 2004). Other important factors, as Wallerstein (1999) or Braudel (1985) point out, are the fragmentation of the social sciences into various detailed disciplines as well as the lack of an interdisciplinary approach. As the aforementioned authors suggest, economy, sociology, anthropology, or history are fields of human knowledge and activity that are not autonomous from one another; they are not based on separate logic so, therefore, they are intellectually coherent. Thus it seems that social issues are best addressed through interdisciplinary study. Such a tendency may be observed especially in reference to works regarding the issues of sustainable development.

First however, the conditions of the social environment should be determined, both material and spiritual. A clean environment and good infrastructure are not enough here. What is the good of them, if the people occupying this space may pose a threat to one another? Furthermore, one of the consequences of lack of respect for other people is the lack of respect for nature.

The scale of negative social phenomena must also be highlighted in the strategies of sustainable development. Counteracting is complicated though, as the social phenomena are complex and determined by various causes. Let three examples from the past serve as a warning.

- Underestimating environmental conditions was characteristic of the Sumerian civilization (Ponting, 1993). This earliest, literate, and advanced civilization (e.g. they used vehicles moving on wheels) had stabilized around 3000 B.C. The land they occupied, located between the rivers Euphrates and Tigris (Lower Mesopotamia), favoured the development of agriculture. The yields were high thanks to the advanced irrigation system. A large increase in population was observed, which consequently entailed an increase in demand for food. After about 2400 B.C., the yield increase had slowed down, then it started decreasing until, around 1800 B.C., it dropped to one-third of that obtained in the prime period. This negative process continued. The signs of the crisis were ignored however, which led to a total collapse of agriculture. Of the many reasons for the yield drop, two deserve special stress:

  - Widespread irrigation had led to an increase in the level of groundwater and washed the salts out of deeper layers of the ground to those closer to the surface; the result: increased soil salinity (one of the major causes of soil degradation).
  - Increased demand for food, together with the growing population, forced expansion of the cultivated area. After the available fields were destroyed, more and more forests were cut down. The area obtained this way was then cultivated. This resulted in soil erosion—another major cause of soil degradation. Moreover, the loss of vegetation, as well as the erosion, made way for the creation of large runoffs and the silting of rivers, which, in turn, caused floods.

Neglecting the environmental consequences of human actions had led to the decline of the entire civilization.

- An environmental factor, lead poisoning, is also pointed to as one of the causes for the fall of ancient Rome. This contamination may have influenced the deterioration of social relations. The Romans used dishes containing lead. The bodies of those who had died in Herculaneum usually contained 84 mg of lead per 1 kg of bone. This statistic does not reflect that the skeletons of some individuals contained over 2000 mg of lead per 1 kg of bone. For the sake of comparison, the skeletons of prehistoric people living in today's Greece contained just 3 mg of lead per 1 kg of bone (Borkowski, 1992).
- Combining social and environmental aspects was also the cause of the decline of the Polynesians in the Easter Islands. What they have left behind were the famous stone statues and a devastated environment. The social structure was a determining agent there. Polynesians formed rival clans and erecting stone statues was the most visible manifestation of this rivalry. The scale of the

sculptures still impresses us today, though there is an enormous technological gap between the time of their formation and our own. It is not only a matter of working the stone, but also the purely technical issue of transporting the blocks of stone from quarries to their placement sites. Working animals would have been a great help, but none were available on the islands. Therefore, in order to move the stones from place to place, they would cut down forests to make trails thus making it easier to shift the blocks. The rivalry between the tribes entailed erecting more and more monoliths. Over time, the wooden trails degraded. Where islands are concerned, wood stocks are always strictly limited. In light of this, the Polynesians ceased building wooden houses and moved to caves, because every tree was needed. Moreover, the fishery declined, since the only known way to construct a boat was to make it out of wood, which was a strategic resource in the system. But this was being used for a different purpose. Thus, one of the major food sources become unavailable. What was left was the traditional agriculture, which, in these specific circumstances, was limited to potato cultivation. After the forests had been cut down, soil erosion became a problem, which had a negative impact on yields. No population can maintain itself without a food supply, and in this case the food ran out. The disaster must have been sudden and unexpected, since many unfinished statues remain near the quarries to this day (Sörlin, 2003). The people died because they had destroyed their environment.

The most important conclusion drawn from the above examples is this: since it is possible to destroy the environment locally, it is also possible to destroy the environment of the whole planet—including the entire biosphere (Douglas, 2006). This is a warning in the context of today's technological powers and of global environmental issues. It is also worth noting that each and every civilization created by humans (except for the present one—so far, at least) collapsed sooner or later, regardless of its cultural or social systems. This is a major challenge for human habits and behavior patterns. This challenge is all the more important, since modern civilization—the scientific-technical one—does not take into account the conditions of the natural environment and is thus making the very same mistake that caused the spectacular social disasters of the past.

The essence of the problem is the need to overcome the existing opposition between humans and nature. This is very deep-rooted in human culture, and has been, paradoxically, strengthened by the contemporary environmental protection and nature conservation.

The first of the two has gained a strong technical character, providing answers to the question of how to clean the environment. Whereas nature conservation tends, as part of the activity described in the previous chapter of this work, to create secure 'isles' separated from humans under strict protection. Undoubtedly, environmentally valuable areas should be protected. Unfortunately, the act of creating even the largest National Park, reservation, or the NATURA 2000 network area is not good enough. That is because the challenge lies in the global character of human operation, which brings about hazards of similar magnitude. Such phenomena as climate change are a threat to every part of the globe, irrespective of how valuable it is in terms of the environment. At the regional level too, air pollution from various industrial chimneys is transported over large distances and creates a problem for unindustrialized areas, including sites that are legally protected.

That is why the discussion on sustainable development assumes a wider approach. In the social aspect, it is a return to the concepts of connecting humans with nature, definitely making them closer than is the present case. We should protect both natural monuments and cultural monuments (Pawłowski, 2002). This corresponds to the purely practical domain, since both groups of objects would intertwine anyway. A particular example is given in the case of the old wooden churches that were erected in deliberately chosen places, important for local landscapes, like hills. What is important is that trees were usually planted around them. Over the years, the whole of the complex would gain status of a monument—the temple would become a cultural monument, while the trees around them would become a natural monuments. This is also recognition of our ancestors, who had intuitionally erected their structures in harmony with the natural landscape.

Thus it is no coincidence that already in 1933 Carl Troll divided the landscape not only into natural elements like geosphere and biosphere, but also distinguished noosphere—that is, the sphere of human thought, which is associated with culture (Rychling, 2004). As Zbigniew Myczkowski notes, culture and nature are inseparable components of our contemporary civilization and joint protection of the two is a determinant of the level of our civilization (Myczkowski, 2001).

2.2 *Cultural landscape*

Culture may be described in many ways. Clyde Klukhohn identifies as many as 120 definitions, which can be divided in six major groups—see Table 25. The early definition of culture dates back to the turn of the 18th and 19th centuries.

Table 25.   Definitions of culture (Adamek, 2001).

Definition type	Range of content
Descriptive/ enumerative	Culture perceived as being identical to the notion of civilization. A complex aggregation of knowledge, beliefs, arts, customs, morality, laws, and human capacity.
Historical	Emphasis on the past, tradition (or to put it more widely—the achievements of human activity) as culture's most important agent.
Normative	Culture as a way of behavior consistent with the previously accepted standards.
Psychological	Culture as all that which humans have accomplished through the process of learning and adaptation to the changing situations.
Structural	Focusing on particular types of culture in specific periods (e.g. medieval culture) and analyzing relations in the created frames. Emphasis is put on material/technical elements, social elements, ideological elements, and mental elements (emotions and attitudes).
Genetic	The subject of analysis is the process of the formation of culture, derived from human nature (the word nature being understood widely here, also as nature, or the essence of a particular thing). This group also holds semiotic definitions, which regard the forms of communication and expression within a system of signs.

Johann C. Herder described it as a typically human tool of adaptation to the environment, due to which humans remain superior to the world of nature. As Bronislaw Malinowski wrote 100 years later, this tool must be partly human and partly spiritual (Dyczewski, 1995). Among the contemporary statements on the subject, John Paul II's voice stands out: "Man always exists in a particular culture, which in turn creates a special relation between people, thus determining the interpersonal and social character of human existence. Culture is all that which makes us more human" (John Paul II, 1980).

According to the classical encyclopedia definition, culture "is a set of principles, rules and ways of human behavior, products of human labor and creativity, a collective achievement of human societies, arising on the ground of particular human biological and social traits and of our living conditions, which is developing and transforming within the historical process. [It is also] a particular stage in the process of the development of societies, which is expressed with the level of control over the forces of nature, with the achieved level of knowledge and arts as well as the forms of social coexistence" (Universal Encyclopedia, 1984) It is, therefore, a form of interpersonal communication (Hall, 1987). Reference to the environment should be highlighted. On the one hand, this includes the obvious biological factors of human existence (Wolanski & Siniarska, 2002; Steward, 2006; Piatek, 2007). On the other, is the issue of coming into contact with the environment; the evolutionary adaptation of humans to their surroundings, which had transformed over time into taking control of nature.

Consistently, the cultural landscape includes the material effects of human activity within the natural environment (Mysliwski, 1999), or even the effect of integration between humans and their natural environment in time and space (Dorste et al., 1995; Richling, 2004)—see Table 26.

Cultural landscape refers to a given space and is shaped individually for every region (Knercer, 2003). The combination of culture and nature strengthens identification of local residents with the particular area (Pawlowska, 2001), thus favoring nature conservation; maybe even better than traditional area protection. There are many papers suggesting that the protection of cultural heritage may be just as important for human development as the protection of nature or economic issues, and that the ecological crisis may be partly the result of a cultural crisis (Prandecka, 1991).

Cultural landscape (Dyczewski, 1995; Hall, 2001) is also a carrier for certain values (a cultural value is described as one that is socially sanctioned within a particular culture, and that helps to make choices, indicates purpose, and means of action). In other words, the material culture related to a cultural landscape is based on spiritual culture. As A. Michalowski (2002) notes, for their spiritual and existential needs people should identify with their place of residence, dwelling, town, or region, should cultivate family traditions, local traditions, and folklore. They should find their regional identity in the surrounding landscape. This is a major factor in determining the high quality of life (Tweed & Sutherland, 2008). Unfortunately, many individual objects (both sacral and secular), traditions, and customs do fall into oblivion.

Perhaps the earliest recorded regulation on the protection of monuments as such, dates back to 222 A.D. It was Emperor Severus Alexander's decree which prohibited the devastation of monuments. Increased interest in the issue was noted in the 18th century and was associated with the discovery of the cities

Table 26. Selected definitions of cultural landscape (Lapinski, 2007, changed).

No.	Authors	Definition
1.	German school of geography at the turn of the 19th and 20th century (e.g. R. Gradman, O. Schluter, J. Wimmer)	Objects (natural and human alike) available to direct perception. The landscape is a fragment of our natural and anthropogenic environment.
2.	German school of geography in the 1920s (e.g. A. Hettner)	Physiognomic expression of the culture of a nation or social group, etc. Not only does it include our material environment, but also spritual elements of human activity.
3.	C. Sauer, USA	A given area with specific relations between natural and cultural elements that are evidence of human existence on Earth throughout the centuries. It is a result of the influence from the local societies and the local natural conditions.
4.	P. Vidal de la Blach, France	It is the result of a social relation between humans and the environment. This provides information on how individual groups interpret, assess, and use their natural environment.
5.	G. Mysliwski, Poland	Material effects of human activity.
6.	Zbigniew Myczkowski, Poland	The traditional field of history, arising from human activity, the appearance of which is evidence for the development of civilization at the national, regional, and local levels.

of Pompeii and Herculaneum, buried in the ashes from the eruption of Mount Vesuvius (Trzeciak, 1957).

It is worth noting in this context the interesting book of Wladyslaw Kozicki published in 1913. Almost 100 years ago, he pointed out that, "we witness a very sad and distressing phenomenon. The authorities and institutions appointed to watch over the monuments are overwhelmed with requests (...) for permission for dismantling of old, venerable and yet so homely and close-to-the-heart wooden churches. The reason in general, is that (...) they have become too small (...) and thus must be replaced with bigger—usually made of brick—places of worship. And unfortunately, it happens all too often, that—before the conservation authorities are able to save a precious monument from complete destruction, before they even begin securing its existence, (...) or at least try to (...) photograph it and describe it—the wooden church, often of prime significance for history of art and culture, is long gone, replaced by a new structure, whose artistic value should be condemned" (Kozicki, 1913). Kozicki saw this as evidence of the present generation's detachment from the ancestral traditions. He did suggest that tradition cannot be learned merely by reading books. A person should not only know, but also see and have contact with a true monument, with a real-life remainder of past times. These words are up-to-date even now, both in reference to places of worship and to secular monuments. What is left often decays for lack of funding and lack of interest and still what is nowhere else is being replaced with what is everywhere (Lipinska, 2003). To some extent, such loss may be compared to the extinction of species in the world of nature. The diversity of cultural landscape, however, is based on the fact that not only can it be preserved or restored, but it can be developed creatively as well (Pawlowska & Swaryczewska 2002; Trzaskowska & Sobczak, 2006). It is thus possible 'to live a new life under the old roof', that is, to maintain the functionality of old monuments and areas (Lizewska & Knercer, 2003).

Moreover, the cultural landscape may, under certain circumstances, be reconstructed in a location other than the one it originated in. The culture of the Mennonites may be recalled. They were Christians and formed closed groups (incidentally, the Amish are descended from them). They were driven out of Switzerland and travelled to Prussia, Poland, Uzbekistan, Canada, and Mexico. They would always reproduce their cultural landscape with care, regardless of where their new place of residence was, and maintain it for generations, while remaining exceptionally resistant to the modernization taking place around them (Margul, 1986; Lapinski, 2007).

How are cultural landscapes protected nowadays?

Within the EU, the following documents are of special importance (Council of Europe, 2000):

- "The Pan-European Strategy of Biological and Landscape Diversity"— apart from purely ecological issues, room was made for the promotion of biological/cultural diversity, which combines environmental issues with the residents' regional cultural identity.
- "The European Landscape Convention", which already protects a series of diverse landscapes. These are perceived as a common heritage, contributing to the preservation of local cultures, while positively affecting the quality of life and the feeling of identity of a society.
- CULTURE 2000—a program, which is an equivalent of the NATURE 2000 program. It has been established by the European Parliament and the Council of Europe's decision no. 508 of 2000. Its task is to extract the community's cultural space, promote dialogue, and accentuate the significance of cultural diversity as well as its impact on the European identity. The program also assumes openings for cultures from outside the EU.

The issues of cultural and natural resources are treated with equal importance within the UN as well. One example is the UNESCO "Recommendation concerning the Safeguarding of Beauty and Character of Landscapes and Sites" of December 1962. It stated that modern economic development causes a significant degradation of both natural and cultural landscapes. This provides the right background for the UNESCO List of World Heritage Sites, which was established by the "Convention for the Protection of the World Cultural and Natural Heritage", proclaimed in 1972 and coming into force in 1975. The convention was set two basic goals:

- Harmonious combination of the beauty of nature with the values of human creations.
- Promotion of monuments in order to inspire interest in individual societies.

Objects are placed on the list on the basis of their universal value, internationally, or even globally. There are two lists of criteria: six factors for cultural objects and a separate set of four criteria for natural objects—see Table 27. In 2010 the list included 911 objects—including 704 cultural, 180 natural, and 27 mixed—in 151 countries (WH UNESCO, 2010).

It is also worthwhile to point out the original solutions adopted by individual states. The example of Poland is very interesting. Since 1994 a list of Historical Monuments (HM) has been compiled, signed by the President of the Republic of Poland.

These monuments are objects of great historical, scientific, and artistic value, important for Polish cultural heritage, that are an inspiration for future generations (Roziewicz & Wendlandt, 2004). They include, for example, urban systems, pieces of architecture and construction (including defense structures), cemeteries, archeological monuments, industrial heritage objects, places commemorating

Table 27. Criteria for designating the universal value of objects submitted to the List of World Heritage Sites (WH UNESCO, 2010).

No.	Cultural criteria	Natural criteria
I.	Objects that show signs of creative genius.	Areas representing an important period in Earth's history.
II.	Objects that affected the further development of architecture.	Areas showing evidence of important ecological and biological processes.
III.	Objects that are a unique manifestation of Culture.	Natural areas characterized by outstanding aesthetic value.
IV.	Objects that are an exceptional work of art representing a given period in history.	Areas of special biological diversity (biodiversity).
V.	Objects representing traditional ways of land development.	–
VI.	Objects related to great historical events.	–

Figure 7. Wroclaw: the Centennial Hall (World Heritage and historical monument). Photo by Artur Pawłowski.

Figure 8.   Cracow: Wawel Royal Castle (World Heritage, and historical monument).
Photo by Artur Pawłowski.

Figure 9.   Haczow: wooden church (World Heritage and historical monument). Photo
by Artur Pawłowski.

historical events and cultural landscapes, parks, and gardens. Thus, in this case
as well, natural aspects are found next to cultural values. In the autumn of 2010,
37 monuments were listed, including the Pauline monastery in Czestochowa, the
Wawel Royal Castle in Cracow, the salt-pit in Wieliczka, and the Centennial Hall

Figure 10. Czestochowa: Jasna Gora—Pauline monastery, spiritual capital of Poland (historical monument). Photo by: Artur Pawłowski.

Figure 11. Zamosc: Marketplace with a town hall (World Heritage and historical monument). Photo by Artur Pawłowski.

in Wroclaw. Many of these places are also listed in the UNESCO WH. Some of these objects are shown in the photos.

Such initiatives, like the list of historical monuments, are of great importance for the preservation and popularization of local cultural values. This is consistent with the spirit of the new international tourist programs. What stands out among them, is the project named PICTURE (Proactive Management of the Impact of Cultural Tourism upon Urban Resources and Economies), the aim of which is "to develop a strategic urban governance framework for the sustainable management of cultural tourism within small and medium-sized European cities. This framework will help to establish, evaluate, and benchmark integrated tourism policies at the local level with a view to maximizing the benefits of tourism upon the conservation and enhancement of built heritage diversity and urban quality of life" (Picture, 2010).

In this context, Jeremy Ryfkin's statement is correct. He said that, taking into account the current level of development of civilization, we are now in the middle of a long-term transition from industrial production to the production of cultural goods (Krzysztofek, 2004). Cultural goods are not, however, only inherited from the past. In the present times, it would be wise to point out factors regarding the two main types of environments occupied by humans—the rural environment and the urban one (Sain Marc, 1979).

2.3   *Urbanization and Healthy Cities*

A process of rapid migration from rural areas to cities is now observed on a global scale. As already mentioned in this work, in 2008 the number of people living in cities exceeded that of people living in rural areas; thus, over half the world's population now live in cities. According to the UN Habitat program and the "New Charter of Athens" (Radziejowski, 2006), these cities must be consistent in terms of not only spatial planning, but also economics and society. This is a goal that is hard to achieve. One of the major obstacles is the existence of 'slums'. Even now, in many agglomerations located in developing countries, e.g. India (Krishna-Hensel, 1999), the proportion of people living in these 'bad neighborhoods' may be as high as 50 to 70% of the total population.

Rapid urbanization also concerns countries with rural traditions, such as Poland (Amyo, 1996). According to official statistics, until the 1960s the vast majority of the citizens lived in rural areas. Currently, this proportion is only about 38%. Furthermore, only 16.7% of the rural population finds employment on farms (Mossakowska, 2005). This means that these areas are becoming more and more multifunctional (Hadynska & Hadynski, 2006), where traditional cultivation is only one of many development opportunities (supplemented by service and industrial sectors). This is the case with over 80% of the counties in Poland (Mossakowska, 2005).

The global process of conversion of traditional rural areas into multifunctional areas is escalated by the recently observed trend of people settling in rural areas next to cities, while working in the city itself. Unfortunately, this sometimes results in the creation of chaotically built settlements, which, as described in one of the works, are neither urban, (...) nor rural—they are bland and vague (Trzaskowska & Sobczak, 2006). In other words, rural counties near cities lose their original characters, while not fully transforming into urban areas. This may not only entail the decline of rural landscapes, but also the disappearance of traditional customs and rural cultures.

Moreover, the emerging new suburban areas may be characterized by various functional categories, which show division into areas with the following characteristics (Stola, 1984):

- Dominant external functions, such as industry and architectural engineering, with moderately developed service functions.
- Co-dominant industrial, architectural engineering, and service functions.
- Co-dominant industrial and agricultural functions, with poorly developed services.
- Co-dominant industrial and service functions.
- Co-dominant recreational, service, and agricultural functions.
- Dominant recreation/health resort functions.

These functions partly overlap with the expectations shared by people leaving the city. They mainly regard the following issues (Thomas et al., 2004):

- The need for contact with nature.
- The need to have one's own garden.
- Having a beautiful view.
- Providing cheaper accommodation space.
- Protecting children against the hazards typical of cities.
- The need for peace and quiet.

The listed expectations have their ecological dimension (contact with nature), economic dimension (cheaper accommodation space), and also social dimension (sense of security, escape from the hustle and bustle of city life). Moreover, one can point to the low population density in rural areas. This prompts the development of strong social relations; everybody knows each other, no one is anonymous.

Things look different in the case of cities. Here, the population is very dense, and the social relations are not as strong as in the country (related to, for example, the high level of anonymity).

Proper, and hence, sustainable, urban environment management is a major challenge in this context. Table 28 presents the classic definitions on this matter.

The discussion was expanded within the last few years. Wojciech Peski (1999) perceives such management as an activity directed at securing the efficient operation and sustainable development of a historically shaped settlement. This general statement is then fleshed out in more detail by indicating the primary objectives of such development:

- To increase the income of both the city and its residents.
- To increase employment.
- To support actions directed at increasing the level of a city's economic self-reliance.
- To support actions directed at improving the state of the environment (Peski, 1999).

These tasks and needs must be realized in such a way as not to impair the future generation's ability to meet their needs—which is a direct reference to sustainable development. Such an approach expands the discussion by environmental and social aspects.

The concepts of 'ecological cities' also goes in this direction (Bithas & Christofakis, 2006). They include the idea of a Responsible City (Andersson et al., 1997), an Eco-polis (Tjallingii, 1995), a Sustainable City (part of a wider concept of the European Conference on Sustainable Cities and Towns, implemented since 1994) or a Healthy City (Anderson et al., 1997). The latter, formulated by the WHO, is the most commonly known.

Table 28.   Urban management—selected definitions (Peski, 1999).

No.	Definition
1.	Business requiring such activity, which provides a more efficient and economic use of loans to the city.
2.	A more effective way of governing a city.
3.	Realization of a policy, i.e. managing a city in the public interest.
4.	Implementing a policy. More than management, since it excludes the private sector, or plays an active function in enterprises aimed at urban development.
5.	Stimulating the process of including governments of individual states into the programs of the World Bank and the UN in terms of urban development.
6.	An activity strictly related to urban planning.
7.	Implementation of urban development programs in a more effective way.
8.	Managing development or growth based on meeting the needs of cities that develop and reconstruct according to their policies on managing land, buildings, and services network.
9.	Control over a city's finances or the state of the environment.
10.	Administration within the local authority's primary responsibilities in terms of providing residents with high level services.
11.	Taking up continuous (long-term) responsibility for activities aimed at performing specific tasks regarding human settlements.

The Healthy Cities Project was initiated in 1986, in 11 locations at first; now about 90 cities are members of the WHO European Healthy Cities Network, and 30 national Healthy Cities networks across the WHO European Region have more than 1400 cities and towns as members (WHO, 2010). It includes a wide variety of urban areas, from little towns to metropolises, from healthy resorts to cities in heavily-industrialized parts of the country (usually suffering from higher air pollution levels, for example). It should be pointed out that every city must make a formal application to be added to the list and, upon its acceptance, must realize the WHO recommendations. These refer to the general definition of a Healthy City, expanded by the list of strategic goals—see Table 29.

According to the WHO definition, a Healthy City "has a clean, safe physical environment, and its development rests on a sustainable ecosystem. It provides save supplies of food, water, and energy, and efficient waste disposal. It also provides entertainment and leisure activities that facilitate interaction and communication among its citizens. A healthy city values its past and respects the diverse cultural heritage and specialities of its citizens regardless of race and religion" (Anderson et al., 1997). The opinions of the residents themselves are also of importance (Tweed & Sutherland, 2007).

Simply put, we may say that this definition consists of three primary objectives, reaching far beyond purely health issues:

- Clean environment (environmental indicators include air pollution level, the quality of water, or health indicators, such as mortality, low weight of infants, and in a wider aspect, the number of citizens per doctor, or the percentage of children undergoing periodic vaccination).

Table 29.   The strategic goals of WHO European Healthy Cities Network (WHO, 2010).

No.	Goal
1.	To promote policies and action for health and sustainable development at the local level and across the WHO European Region,with an emphasis on the determinants of health, people living in poverty and the needs of vulnerable groups.
2.	To strengthen the national standing of Healthy Cities in the context of policies for health development, public health and urban regeneration with emphasis on national-local cooperation.
3.	To generate policy and practice expertise, good evidence, knowledge and methods that can be used to promote health in all cities in the Region.
4.	To promote solidarity, cooperation and working links between European cities and networks and with cities and networks participating in the Healthy Cities movement.
5.	To play an active role in advocating for health at the European and global levels through partnerships with other agencies concerned with urban issues and networks of local authorities.
6.	To increase the accessibility of the WHO European Network to all Member States in the European Region.

- Good infrastructure (chosen indicators include transport, healthcare, employment).
- Good socio-economic conditions (chosen indicators are the number of homeless people, the number of unemployed, the crime rate, as well as good social relations between people).

This part of the work focuses on the third pillar, that is, the social issues. This is because urban centers may be perceived as a social system, in which the community puts out a stable effort to occupy, manage, and control a specific area (Chojnicki, 1988).

The issue of increasing hostility and aggression between residents is a particular problem here. Such attitudes are the consequence of the struggle for existence, as well as, as a characteristic of our time; striving to take the best possible place in the social hierarchy. According to Wieslaw Sztumski, the greatest level of violence in history has been reached in modern times, when all human activity became subject to the pursuit of profit and to the paradigm of economic rationality, efficiency of income, and political efficiency (Sztumski, 2000). This is accompanied by the rejection of previous ethical systems. A lack of restrictions allows for the development of a consumerist model of life. Only a few win the 'race to welfare', which results in the formation of individualistic and egoistic attitudes on the one hand, and the rise of further social conflicts on the other. Those who are not sufficiently lucky to gain welfare, envy those 'chosen ones' who do. This is associated with a large amount of aggression. The literature even has a term for that—social exclusion[4], which means

---

[4] The notion of 'social exclusion' was formulated in France, in reference to people who were not granted social security. Nowadays, the term's meaning is much wider. Structural marginalization may refer to individuals, or large social groups (even residents of a particular village or town), but may also refer to the gap between the rich and the poor countries (Gore, 1995).

the structural marginalization of people. This combines social issues with economic conditions, the major factor being a high level of long-term structural unemployment. However, in places where there are still jobs available, the people have so little time for one another that the weakening of interpersonal relations, family relations or even those relations regarding one's identity with a specific region or social group—takes place.

The issue of aggression is very strongly marked in the countries of Eastern Europe within the residential areas of so-called 'tower blocks'. These concrete blocks, 10 or more floors high, with usually small apartments, were supposed to solve the problems of a housing shortage, as well as be cheap yet solid. As it turned out, they brought about social problems over time, including increasing violence, especially among youth.

Tower blocks are sometimes described as 'urban cracks' (Skalski, 2007), because, as far as the city's character is multifunctional, these dwellings are purely playing the role of 'bedrooms'. Other important functions, such as services, commerce, or socio-cultural ones, are underdeveloped or even non-existent. This causes the formation of a sort of barrier not only socially speaking, but also in terms of spatial planning. This separates the bedrooms: from the rest of the city, which is usually perceived as more attractive.

The following consequences are very important (Skalski, 2007; Banka, 1993):

- An unstable life situation.
- Relations restricted to the closest family.
- The sense of 'social exclusion'.
- Increased sense of insecurity.
- Increased violence and a higher crime rate, mainly among the youth.
- Reduced activity for the residents, hence the difficulty of mobilizing them to any action in their neighborhood.

Every one of these problems becomes another barrier. Somatic disorders take place as well, such as the sick building syndrome (SBS), which causes fatigue, nervous system disorders, dyspnea, and a low level of concentration (Mizielinski, 2002).

These phenomena require programs that improve infrastructure, both technically and socially. Radical action is taken at times, e.g. tower blocks are being dismantled in many countries, sometimes leaving only the lower buildings (up to 5 floors high), and even those are described as 'low-standard housing' (Peski, 1999).

The fundamental challenge will be the problem of improving the situation of the people living in the existing estates. Part of the implemented programs is restricted to technical infrastructure only. This was the case in the former GDR, German Democratic Republic (Skalski, 2007). The problem, however, was to find people willing to live in these rebuilt, renovated, but also more expensive, blocks. Wealthy citizens were not interested in such flats, while the poorer people, if only for financial reasons, tended to stay in the old housing. Apartments which no one wants to live in are definitely not sustainable (Greiff, 2005).

Taking into account the social plane of sustainable development can provide good results. Some papers mention the necessity for 'social space'—infrastructure elements, such as clubs, common-rooms, bars, or other places, where residents meet each other and spend their free time together. With attractive forms of activity within these units, which do, however, require financial backing, the relations between neighbors improve, the residents become more active, and the crime rate drops (Zaborska, 2007).

Unfortunately, in the case of 'tower blocks', even if there is such a 'social space' facility within the estate, their budgets are usually low and their operation very limited. In this situation, people who can afford it are moving to better estates with higher standards, usually fenced off and guarded (Zaborska, 2007). This makes the gap between former neighbors even deeper, and puts yet another obstacle in place, which strengthens the opposition between the rich and the poor. This opposition is marked at the global level as well: rich North on the one side, poor South on the other.

2.4 *North vs. South*

We may divide the world into few groups, as Immanuel Wallerstein (2004) did:

- Rich 'core' countries.
- Poor 'periphery' countries placed at the other end of the scale.
- 'Semi-peripheral' countries, which might advance to the core, but if they fail to use wisely their opportunity, then they would share the fate of peripheral countries.

The most important issue is the opposition between the first two groups, which represent the countries of the rich North and of the poor South. Despite the many UN programs intended to bridge the gap between these two groups of countries, it is still growing and this seems to one of the major problems of our time (Kozo, 1994; Greig et al., 2007).

It must be noted that, from the environmental aspect, the rich countries have had a period of ruthless devastation of the environment. Currently, they are requiring the poor countries not to destroy theirs. Meanwhile, their help for the South is very limited. In such a situation, are the countries of the South really obliged to accept this sacrifice for the environment? Do they have a chance of avoiding the North's past mistakes? Moreover, is the North ready to introduce more radical changes within it?

These are major questions, since human pressure on the environment and its resources is growing continuously (by as much as 50% between 1950 and 2000). The statistical data is meaningful; just one-fifth of the world's population lives in the North, but it consumes up to 70% of the total energy produced, 75% of the metals, and 85% of the wood (Schnoor, 2003). These resources are squandered to an alarming extent: research shows that 93% of resources used in the USA are not processed in any way into commercial goods. Furthermore, 80% of the ready-made goods are thrown away after a single use, while the rest are not as durable as they should be (Weizsäcker et al., 1999).

It should be highlighted that the deposits of most of the resources the North squanders, are located in the South, and their exploitation is the main cause of environmental degradation in the poor countries. Taking into account the fact that it is physically impossible for all humanity to consume resources at the pace of the rich countries, it seems right to postulate that these very countries should undertake the most radical steps (Bouce, 1996). Present initiatives are not sufficient; impoverished areas are growing continuously. The gap between the richest and the poorest in 1830 was expressed as a 3:1 ratio; in 1992 the ratio was 72:1.

The income of the poorest countries in the world is now only 1.9% of that of the richest ones (Wise, 2001). The level of poverty in Latin America was 13% in 1985 and reached 52% a few years later (Esty & Gentry, 1997). The level of consumption in an average African home between 1973 and 1998 decreased by as much as 20% (Clark, 2007). At present, over a billion people have an income of less than US $1 per day, and almost three billion people earn no more than US $ 2 a day (Wise, 2001)[5]. Figure 12 shows the spectacular economic increase in the richest countries, together with the poverty of the poorest countries of the South.

The poverty of the South is not just economically driven but is also structural.

Let us point out several basic issues (Heredia, 1996):

- Lack of democracy in many countries and thus no control over the aid, which, instead of going to those in need, was divided between a few prominent people.
- Carrying out policies favoring monopolies.
- Low government capacity to provide the basic social services, including education.
- Lack of actual access of societies to production means and resources.
- Lack of an appropriate distribution of goods.
- Focusing the national economy on Western markets rather than on the country's own needs, which has resulted in debts for the poor countries and the need to pay them.
- Overexploitation of natural resources and environmental degradation.

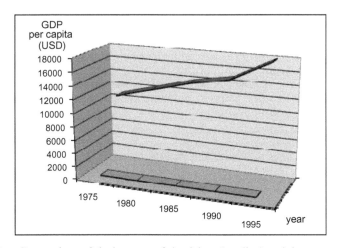

Figure 12.   Comparison of the incomes of the richest (top line) and the poorest (bottom line) countries (Harris et al., 2001, changed).

---

[5] See some interesting analyses of the situation in selected developing countries: Chile, Pakistan, Tunis, Turkey, Zimbabwe and the ever more prosperous China (Luken, 2007) as well as an article comparing the opportunities for development in 18 developing countries (Luken, 2006); an essay comparing the situation in almost 30 rich and poor countries (Nagpal & Foltz, 1995) and a detailed article on the situation in Nigeria (Ite, 2007).

The issue of poverty is also related to the demographic crisis (Eckholm, 1982; Mackenzie & Mackenzie, 1995; Glasby, 1995), which includes the hazards of a rapidly increasing population. As scientists of the Sierra Club warn, no matter how well the implementation of new development strategies goes, the rapid increase in population may ruin all the efforts (Sierra Club, 2010; Marten, 2001).

Let us present the available data. During the Neolithic age, there were about 10 million human beings. At the beginning of our era, there were about 300 million, by 1250 this number had increased to about 500 million, and by 1800 the population stood at 1.75 billion (Bratkowski, 1991). Even in 1900, the annual rate of population growth was around 1%. In the 20th century, the highest level (2.2%) was reached in 1964. The total human population had increased fivefold between 1820 and 1992 (Streeten, 1988).

In 2010, the total number of people was 6.8 billion, having increased by 800 million since the year 2000 (World Population ..., 2010). The number is increasing by about 80 million per year, 9132 per hour, 152 per minute; 98% of the increase is occurring in the developing countries (Pinstrup-Andersen & Pandaya-Lorch, 1998). Currently 75% of the global population (almost 5 billion people) lives within societies in developing countries—see Figure 13. The population of some particularly populous countries, such as India and China, may triple before 2050 (Goodlan, 1995; Kozlowski, 2005)—see Table 30.

The situation is expected to stabilize in 2075 at around of 9.22 billion people, which is almost twice the population of 2000 (World Population ..., 2004).

Uneven population increases in different parts of the world does not mean that the consequences will only affect some chosen regions. It is, therefore, a global issue. Millions migrants from the poor countries are absorbed into the rich countries every year, while half of the population increase in the USA occurs through immigration (Harris, 2001). The rich North is not an isolated island; even the wealthiest are now experiencing a sense of discomfort. The issues of aggression, threat, or terrorism apply to every human being, regardless of their social or material status. This was best shown in the terrorist attacks of September 11, 2001. It might be worthwhile to summarize B.R. Barber (1995) who perceives the modern world as an area, where two wars are being fought.

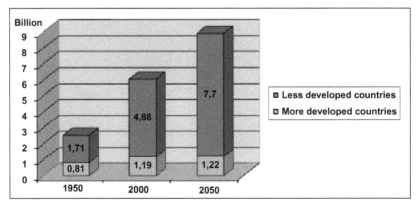

Figure 13.   Present and forecasted changes in human population (World Population ..., 2004, changed).

Table 30. Top 10 countries by population and forecasted top 10 by 2050 (2010 World Population Data Sheet, 2011).

2007			2050 (forecast)		
No.	Country	Population (million)	No.	Country	Population (million)
1.	China	1,338	1.	India	1,748
2.	India	1,189	2.	China	1,437
3.	USA	310	3.	USA	423
4.	Indonesia	235	4.	Pakistan	335
5.	Brazil	193	5.	Nigeria	326
6.	Pakistan	185	6.	Indonesia	309
7.	Bangladesh	164	7.	Bangladesh	222
8.	Nigeria	158	8.	Brazil	215
9.	Russia	142	9.	Ethiopia	174
10.	Japan	127	10.	Congo	166

The first of them is a bloodless struggle for profit, fought by the wealthy against the poor; the second is a symbolic Jihad, representing the violent 'bloody identity policy', which expresses opposition to the values and wealth of the North.

It is worth to ask: is the demographic problem exaggerated? As mentioned before, stabilization of the increase will be possible by 2075. This is not a particularly distant prospect, hence another question: if the demographic problem is exaggerated, is the proposal for a radical reduction in this increase, which appear in numerous development strategies, justified?

It needs to be highlighted that the said proposal not only answers some of the ecological and social issues of the modern world, but also forms a specific development of the doctrine, introduced by Thomas R. Malthus[6] in 1798. According to Malthus, humanity is characterized by a tendency to increase its numbers at a rate higher than the rate of increase in the food supply. Therefore, it is necessary to restrict the rate of population growth artificially (Kozak, 1993). Although Malthus had underestimated technological progress, as well as the developments in agriculture, his ideas are still being discussed. A similar context can be found in China's demographic policy, where parents are allowed to have only one child. The situation in other countries however shows how this can be achieved by more humanitarian means—by increasing the level of education (particularly among women), as well as other basic services, such as healthcare (Harris, 2001). This is a transition from a state of high birth rate and low child survival to a low birth rate and high child survival.

Paradoxically, the problem also appears when the birth rate is too low. Such a situation is observed even now in the rich countries (as well as in the

---

[6] Thomas Robert Malthus (1766–1834)—a British Professor of Economy, a representative of the classical school of thought in economy as well as a demographer and an Anglican priest. Member of the Royal Society, London, the French Academy of Moral and Political Sciences and the Academy of Sciences in Berlin. The supporters of Malthusianism demanded the adoption of 'moral restraint'. This meant that the poor should refrain from having children. What needs pointing out, is that this was, by definition, not a voluntary postulate (Wikipedia, 2011).

post-communist countries), where the birth rate oscillates around (or below) the population replacement rate. This results in the phenomenon of an aging population, which is also a challenge for public finances.

Therefore, demographic issues cannot be absolute, the major problem turns out to be the unequal distribution of food, not its shortage. Moreover, as Jay W. Forrester (1971) notes, even a radical decrease in human population may not succeed in changing the global situation, unless significant change is introduced at the socio-economic level. The resources are mainly consumed by the rich countries, to meet their industrial needs. The populations in those countries undergo a stable reduction, which, nevertheless, does not entail lowering the pressure on the environment. In this context, industrialization may prove to be a bigger problem than the growing population.

That is why demography is usually placed in a wider context when creating indicator systems, such as the Human Development Index, or the Human Poverty Index (Wise, 2001; Skowronski, 2008). These indexes consider three major problem groups:

- Health issues.
- The level of a society's education.
- Economic issues (including GDP).

Taking into account the gap between the North and the South in terms of each of these issues, a proposal has been put forward to introduce separate detailed indexes for each of these two groups of countries. These new indices would report on the following:

- Percentage of people who would live to their forties in a given society.
- Level of education, including illiteracy among adults.
- General socio-economic indexes: e.g. the percentage of starving children, percentage of people with no access to clean water, or even basic healthcare.

For the countries of the North, these standards are higher. Life expectancy moves to the age of 70 and the third factor would include such indicators as the percentage of people with low income and the level of long-lasting unemployment (Wise, 2001). That is because access to clean water and basic healthcare are almost fully provided in those countries.

Furthermore, it is difficult to compare levels of education. After all, the widespread illiteracy in many poor countries restricts the opportunities for further development of whole societies. The problem is particularly visible in Africa, where, according to UNICEF, as many as 862 million adults cannot read or write, and over 120 million children (there are countries, like Liberia, where 90% of children are illiterate) are given no education whatsoever. Two-thirds of children deprived of education are girls.

In this context, it is significant, that such a well-known institution as the Intergovernmental Panel on Climate Change (IPCC) has estimated the worth of human life in rich countries at 15 times that of people living in poor countries (IPCC, 2007).

Moreover, investors from the rich countries very often make use of lower environmental standards in the poor countries of the South, sometimes even pressing for further reductions of such standards, threatening to withdraw their capital otherwise.

This is a practical part of the NIMBY (not in my backyard) principle (Runc, 1988) and 'ecological dumping' (Kosmicki, 1993). For instance, the locals in Ecuador put up a fight against Texaco, demanding compensation for environmental (e.g. leakages from installations) and social damage. This was estimated at US $ 1.5 billion. Suddenly, the government of Ecuador, which, in theory, represents its citizens, began efforts to settle the agreement, which would lead to the payment of only US $ 15 million (Martinez-Alier, 2001). The reason for this was the difference in the adopted standards. The first amount had been calculated according to procedures in force in the USA, the second, according to ones in force in Ecuador.

Can this state of affairs be changed?

Yes it can, although this type of action is still rare and taken in a chaotic manner.

A positive example can be found in Europe, in reference to trade in bananas with Costa Rica. An increase in standards in this country was demanded by European consumers, which was interpreted as the result of a broad-based ecological education (Esty & Gentry, 1997). This also serves as an example that not only manufacturers, but consumers as well, are a group which can actually affect the solutions adopted.

This influence, however, has not always been consistent with the principle of sustainable development. Let us mention the visible resistance of many local societies to actions for environmental protection in reference to the creation of new protected areas (especially national parks and nature reserves).

Species protection may raise controversy as well. The protection of elephants in Africa may serve as an example. These animals were killed en masse, not for food, but for their precious tusks. The protection of this species has been set as a priority, but the methods of its realization are different in different countries.

Kenya adopted a strict law, introducing a ban on hunting these animals and the extraction or sale of ivory became illegal. Unfortunately, these regulations failed to stop poachers, despite the destruction of confiscated specimens on multiple occasions. According to experts, this results from the social background. Individual tribes perceive elephants as a common good which has no owner. Thus, the restrictions were perceived as being directed against local communities. In this new situation, the locals determined that poaching elephants, if successful, is a better solution than protection, as people could then enjoy some of the benefits of which they had been deprived.

At the same time, in Botswana, the management of elephants was handed over to the tribes, and a permit for limited hunting was issued. In such a situation, the tribes proved to be the elephants' best guardians, since the depletion of their population would mean a significant drop in tribal income. The elephants became the property of the local communities, and people usually care for their property (Kalinowska, 2003; Fothergill, 2006).

This example shows that the suggestion that people of the South, if left to themselves, would quickly cause a rapid devastation of their environment, is not necessarily true. Under proper circumstances, they would care for it, seeing how their own survival depends on it. Therefore, the tendency, shared by some ecological organizations, to neglect the rights of local communities (not only in Africa) to co-decide on the ways of managing their land, and to demand its full protection, should be assessed negatively. While establishing new national parks in the era of sustainable development, local traditions and ways of management should be taken also into account. Much depends on proper ecological education, which should not only show the natural value of a particular area, but also the benefits resulting from protecting it, e.g. regarding the development of the tourism sector. Already the local communities themselves are seeking to create protected areas in many places.

Much can be achieved with a wider regional agreement. An interesting example is given in Eastern Europe: the Green Lungs of Poland—Zielone Pluca Polski (ZPP), which covers 18% of the country and includes many legally protected areas (Dolega & Siedlecka-Siwuda, 2006).

The idea was originally introduced in 1983.

According to its author, Krzysztof Wolfram, the uniqueness of the ZPP area is mainly expressed by the large amount and good state of its natural environment

Table 31.   Strategic and operational objectives of the Green Lungs of Poland (Lenart, 2008, modified).

No.	Objective
1.	Increase competitiveness of the region in Poland and in Europe by: creating conditions for an increase of competitiveness of an area of special natural, landscape and cultural values, which facilitate development of ecological agriculture and tourism; enabling use of the area as a unique, model zone of ecological safety in the context of the national and international scale; establishing a friendly and efficient international cooperation, with particular consideration for cross-border cooperation.
2.	Guide pro-ecologically the region's socio-economic development by: stimulating and supporting actions to create opportunities for the economic development of the residents with no harm to the environment; introducing the principle of sustainable development in the projected strategic documents on a local scale.
3.	Strengthen the values of natural and cultural environment by: stimulating and coordinating activities to create new protected areas; intensifying forestation; educating people on the value of protecting natural and cultural variety and its heritage.
4.	Ensure cohesion in the area's diversity, identity, and positive image by: establishing cooperation between regional and local authorities, government administration, cultural and scientific societies, and regional and local media; taking actions that promote residents' identification with the idea of ZPP.
5.	Improvement and promotion of ZPP's 'regional product' by: inspiring and coordinating actions in the field of territorial marketing, facilitating ZPP's development; introducing ecological and spatial standards to the region.
6.	Consideration of the area in the country's spatial and regional policy through: actions for a common regional policy of central and local governments for the area that holds particularly good conditions for a model realization of the principle of sustainable development; actions to promote regionally specific management of natural resources.
7.	Increase the level of knowledge about sustainable development by: popularizing knowledge of ZPP at all levels of education in terms of ecology, protection of the environment, and sustainable development; organizing research and development activity.

in comparison to the general condition of the natural environment in all of Europe (Wolfram, 2005). The social dimension is also important; the variety of cultures and customs (both in terms of nationality and religion), as well as the fundamental issue of development opportunities for residents. Protection of the region's natural and cultural values includes the realization of a number of strategic and operational objectives, listed in Table 31.

Such solutions as the ZPP are, in fact, practical local strategies of sustainable development. What is important is that they merge different aspects of the issue in question. We may point out, for example, the ecological aspect (nature conservation), the social aspect (engagement of local communities), or the technical aspects (development of the necessary infrastructure). In the end, however, the financial aspect may prove to be the most significant, if only in terms of gathering enough money to fund the realization of the goals. This, however, belongs to the economic plane of sustainable development.

3 ECONOMIC PLANE

3.1 *Traditional economy vs. ecological economics*

Economics (Piontek, 2001; Geise, 2000) is a science of the relationships and laws governing the production, distribution, and exchange of goods at various levels of social development (economic growth). It is also a science of the instruments (and organizational solutions), the implementation of which directly and indirectly affect the economic, social, and natural processes (Reut, 2000; Piontek, 2001). Additionally, it is the science of capitalism, which is presently the dominating system in the world.

We add a note of importance here. The conceptual founders of capitalism, such as Adam Smith[7], or David Ricardo[8], based their reasoning on an ethical background, as well as social and environmental ones. For instance, Smith

---

[7] Adam Smith (1723–1790)—a British economist, teacher of rhetoric, belles-lettres and law, Professor of logic, moral philosophy (lectures on ethics, theology, law, and political economy); Lord Rector, University of Glasgow; Founder Member of the Academy of Sciences, Edinburgh; member of the Glasgow Literary Society, the Philosophy Society of Edinburgh, and the Glasgow Political Economy Club. He was the co-founder of a classical school of thought in economics. He was a supporter of the concept of the natural order and liberalism, and an opponent of monopolies. He had formulated the basic principles for analyzing market mechanisms (pricing system, the mechanism of 'invisible hand of the market'). According to his works, profit was not the basis of welfare, but rather work was (Universal Encyclopedia, 1984).

[8] David Ricardo (1772–1823)—an English economist, theorist of the classical school in economics. Originally, he dealt with geology (cofounder of the Geological Society of London), mineralogy, and mathematics. As an exceptionally good stock broker, at the age of 25 he became one of the wealthiest people in Great Britain. He developed his interests under A. Smith. These included the theory of money and credit. He argued that the value of products is set by the amount of work needed to make them. He also drew attention to the distribution of the generated value between the social classes (distribution theory). He claimed it was wrong when only landowners benefited from progress in the economy. He also founded the theory of comparative costs, referring to the benefits of specialization in terms of the international division of labor (Universal Encyclopedia, 1984).

argued that what benefits most people cannot pose a problem for the rest. He also stressed the importance of natural resources, joining this with the idea of the common good, which should be protected (Ikerd, 2008).

The neoclassical economists, who now dominate, have omitted these aspects, focusing instead on 'the invisible hand of the market'. This system, sometimes called 'industrial capitalism' (Lovins, 2004), developed over the 19th and 20th centuries along with transformations regarding the concentration of capital, production, and new organization of the market. The countries that profited most were the United States, and in Europe, Great Britain and France. Following World War I, the group was joined by Germany (together with the annexed Austria) and Japan (Mucha-Leszko, 2005).

This period is also marked by the foundation of the Bank for International Settlements in The Hague, which is considered the earliest international financial institution in the world (Wojtas, 2005).

The free market in its final form was fully shaped in the 1940s, on the ashes of World War II (Harris & Goodwin, 2001). This was a time of restoration and then of progress, associated with the establishment of major financial institutions, such as (Harris et al., 2001; Wojtas, 2005):

- The International Monetary Fund (IMF), established in July 1944. Its goal is to watch over the global exchange market.
- The World Bank (made up of two development institutions: the International Bank for Reconstruction and Development and the International Development Association). Operating since June 1946, its goal is to provide means for introducing developmental programs.
- The Organization for European Economic Cooperation (OEEC) established in 1948. In 1961 it was renamed the Organization for Economic Cooperation and Development (OECD).

Also important was the adoption of the UN Charter (Gupta, 2002) in 1945 and the General Agreement on Tariffs and Trade (GATT) on January 1, 1948 (Gallagher, 2001). These documents outlined the global principles of economic cooperation.

The free market was the domain of the Western countries at first, but became almost homogenous world-wide following the fall of communism in 1989. It is symbolic that, when in 1994 the World Trade Organization (WTO), the continuation of GATT, was established, its signatories included countries of Eastern Europe.

Currently the largest trade group on the global scale is the Asia-Pacific Economic Corporation (APEC, both the USA and China are members) which was established in 1993.

Moreover, the USA, together with Japan and the EU, form the 'triad', which realizes the largest financial flows for trade and capital investment.

These countries compete; the competition between the USA and the EU being the most visible. The balance is currently negative for the Community. Why is this?

First of all, this is shown by an analysis of the Gross Domestic Product (GDP) index, which relates to the value of newly-created goods and services in a particular country throughout the year (Holden & Linnerud, 2007; Repetto, 1991). This index has stabilized in the EU and has stood at about 70% of that of the US since the late 1970s (in the 1950s it was the other way around).

Second, performance at work in European countries is still a lot lower than that in the USA; in 2004 it was lower by as much as 40% (Mucha-Leszko, 2005; Czerniak, 2005).

Third, the EU is more an outlet market for the US than it is a supply market. This results in the Union running a trade deficit. Such a deficit also characterizes the Community's trade relations with Japan since 1968 (Pasierbiak & Kuspit, 2005).

The situation may change, however. For instance, in 1974 the USA was the only manufacturer of computer DRAM memory chips in the world. In 1980, its share of the market dropped to 56%, and in 1987 it satisfied only one-fifth of the world demand for these chips—the rest were mainly produced by the Japanese (Orr, 1992).

Currently, the wars in Iraq and Afghanistan are the major threats; their consequences include a serious crisis for the American economy. It is not known yet, whether the crisis will be reversible, or whether the situation will lead to a disaster in the free market, comparable to the 'Big Crash' of 1929.

Changes can be brought by new agreements as well. Discussions are being held for example on the Free Trade Area of the Americas (FTAA). On the one hand, the result of such an agreement would be the emergence of a new balance of power in the global economy and a strengthening of the USA. On the other, we must not underestimate the economic power of China, which has been growing rapidly during the last few years.

No matter what the course of things is, the operative foundation of the global economy is still the paradigm of the free market, which remains controversial. Although in 1992 the World Bank introduced the principle of sustainable development in its documents (Sachs, 1993), the problem remains that free-market left to itself leads us to excessive exploitation of the environment and increasing social conflicts (Andreasson-Gren et al., 1992; Pawłowski, 2008). It is also a system that functions entirely differently in rich countries than in poor ones. Simple implementation of a global free market may not necessarily improve the situation in regions that were previously subject to central planning. Moreover, new threats emerge, ones covering a very broad spectrum, from economic, through political, to social. Several things that should be pointed out are that strong mafia organizations penetrating government structures, the development of unofficial black markets, corruption, political instability and increasing poverty (Dobrzanski, 2001).

Of course, the free market and economic development are important aspects that need to be considered not only in terms of the traditional economy, but also in terms of the discussion on sustainable development. Although in the first case they were not limited in any way, least of all ecologically, in the second case they have been put into a wider context from the very beginning. It is claimed that in the face of environmental degradation, the independence of the economy from the natural sciences has become a luxury we cannot afford anymore (Harris & Goodwin, 2001). It is, therefore, postulated that the natural environment should be included in budget bills, which would mean evaluating the functions performed by it. It is a very difficult task, as it is not easy to answer clearly the following sorts of questions. How much are ecological services (such as the beauty of landscape) worth? Can a market price be set for them? (German Advisory Council, 2001; Marten, 2001). The importance of these questions was proved in 2000, when the UN urged that research be carried out concerning the products and services provided to humans by ecosystems. This was done in order to properly include them in budget bills (Kereiva & Marvier, 2007).

When it comes to traditional economic indicators, such as GDP, these issues are almost completely neglected, mostly because natural resources are usually considered to be unlimited and 'external'. Therefore, although it is better to have a higher GDP, which allows for increased funds for environmental protection, in practice, countries with a high GDP are often unwilling to take such action (Platje, 2006).

Thus, in a traditional budget bill, environmental costs are not fully covered, which means they are externalized (Wallerstein, 2004; MicMichael, 2003; Kozlowski, 1991).

One of the first attempts at a reverse approach, i.e. one that assumes the necessity to internationalize the environment, were the works of Alfred Marshall, particularly the tax proposed by his student Arthur C. Pigou. It assumed the necessity of establishing a fee, which would be equal to the monetary value of the estimated external costs, in which the environment was also included (Albrecht, 2006; Jezowski, 2000).

This proposal, made in 1920, was ahead of its time, but was later explicated within the Polluter Pays Principle (PPP), proposed in 1972 by the countries of the OECD. This may be formulated as follows: the price of a product or service must include the cost of its manufacture and the cost of the resources used, including the resources of the natural environment. Therefore, the goal of this principle is to fully introduce, through such fees, use of the environment into the sphere of economics (Masclet, 1999). In other words, the polluter should bear the costs associated with securing the environment against any possible degradation resulting from his activity (Ryden et al., 2003).

The PPP is one of the pillars of the currently developed ecological economy (other principles are presented in Table 32), which may be defined as a new way of understanding relations between human economic activity and its impact on the environment (Harris & Goodwin, 2001).

The first task is to identify the resources, or more precisely, capitals, which we should take into account in the budget bill (Leva, 2004). These include (Etkins et al., 1992; Goodland & Daly, 1995; Ukidwe & Bakshi, 2005; Lin, 2007):

- Financial capital: including macroeconomic planning and budget management.
- Physical capital: assets related to infrastructure (buildings, roads, and industrial plants, such as power plants).
- Human capital: health, education and the availability of employment.
- Social capital: capabilities possessed by citizens and interpersonal relations—and likewise for social institutions—including the legal system in force.
- Natural capital, which may be renewable or non-renewable (Constanza & Daly, 2001), market-oriented or not. This capital is created by resources and ecological services, including climate, food, and water.

We may simplify this and distinguish two basic capitals: the capital made by humans and the natural one (Baltscheffsky, 1992). These are very strongly interrelated. For instance (Harris, 2001), a fishing boat (manufactured capital) will not serve its purpose if there are no fish in the water (natural capital).

Environmental economy accepts the necessity to express both capitals in monetary values. Two positions are usually adopted (Atkinson et al. 2007).

Table 32.   Economic principles in environmental protection (Ryden et al., 2003, changed).

Principle	Short description
PPP I Polluter Pays Principle	The polluting unit should bear the cost resulting from securing the environment against possible degradation.
PPP II Polluters Pay Principle	This principle refers to all polluting units; the total fee is collected in proportion to the impact each plant in the region has on the environment.
UPP User Pays Principle	The user of a natural resource should pay the full price for the value of the resource they consume.
VPP Victim Pays Principle	In the situation, when for some reason the polluter is unable to pay the full amount for the damage done, the 'victim' may grant the polluter a subsidy. For instance, a significant improvement in the state of the environment in Sweden was achieved by subsidizing pro-environmental investments in Poland (which, due to the wind circulation, was the source of large amounts of air pollution in Nordic countries).
PP Prevention Principle	Avoiding irreversible changes in the environment by introducing safety procedures.
EFP Economic Efficiency Principle	Taking into account the full environmental costs in product prices, thus reducing the number of other interventions necessary to achieve environ- mental goals.
D Decoupling	It refers to the ability of an economy to grow without corresponding increases in environmental pressure.

- Strong sustainability, which assumes preserving both types of capital[9].
- Weak sustainability, which only postulates preserving the total stock of capital resources. Depleting natural capital must be accompanied by the accumulation of man-made capital. It is therefore an exchange of one type of capital for another. The problem is, there are physical barriers to this exchange, e.g. total depletion of resources would render impossible any further growth of human capital. What is more, some resources (e.g. clean water) are described as critical, that is ones that cannot be replaced with capital made by humans (Munda, 1997).

Development based on exploiting resources will last just as long as it is possible to produce substitutes for at least some natural resources and as long the production of these substitutes compensates for the depletion of the natural resources. In other words, sustainability in terms of the economic plane of sustainable development is achieved when the amount of resources used for generating welfare is being constantly reduced to such a size and quality that it would neither cause extensive exploitation of resources, nor cause the ecosphere to overload (Ryden, 1997).

It should be highlighted that implementing only strong sustainability is not possible, since we cannot refrain from acquiring resources completely. But basing

---

[9] Some works also distinguish extremely strong sustainability, which postulates refraining completely from using non-renewable resources, and exploiting renewable ones only to the level of their natural self-restoration (Goodland & Daly, 2001).

development only on weak sustainability is also not sustainable. Using a mixed solution is the key; take a strong position in the case of some goods and take a weak one in the case of others (Spangenberg, 2005). The choice is not easy. One needs to take into account such things as (Daly, 1996):

- Allocation of resources (their use in production processes may be efficient or not).
- Distribution of goods (may be fair or unfair).
- Scale of enterprises in reference to the environment (sustainable or not).

In the case of protection of the environment, the problem is not only evaluating the damage done to the environment, but also evaluating the benefits of refraining from destructive actions (Giordano, 2006).

Another step, after identifying the capitals and the related conditions, is the actual attempt at evaluation, which is particularly important in reference to natural capital. What are the estimated values of services provided by the environment? The discrepancy is immense: from US $ 2.9 trillion (2.9×10^{12}) up to US $ 54 trillion a year (Constanza et al., 1997; Harris, 2001; Ikerd, 2005). These discrepancies result, for example, from the multitude of factors that slip past all assessments and make it harder to estimate the amount to pay for the functions performed by the environment. For instance: it is hard to evaluate species which we will never know, as they were pushed to extinction by human activity (e.g. cutting down rainforests). Still, even 60% of services provided by the environment are subject to degradation or unsustainable forms of management (World Resources Institute, 2005).

The price may also depend on the method in use. The most direct method refers to the situation when—due to environmental degradation—some particular function of nature is no longer fulfilled. Floods resulting from deforestation are one such example. They are conditioned by the type of plant cover in the catchment areas. When a forest grows in the area, it is able to hold back water, even in heavy rainfall. When it is cut down, however, the barrier, which previously held water, goes with it. The water would then quickly flow down to the rivers and give rise to floods. Such association proved critical in the valley of the river Yangtze in China. Systematic deforestation carried out in the catchment area led to a colossal flood in 1988 that killed 3700 people; 60 million acres of crop fields were degraded, and the general losses were estimated to be around US $ 30 billion. This was the trigger which not only led to the quick passing of a moratorium prohibiting further deforestation, but also to the introduction of a massive reforestation program which cost US $ 12 billion (Lovins, 2004). Thus, the total cost was US $ 42 billion, which greatly exceeded the anticipated profit from deforestation.

Environmental evaluation need not be carried out only in the face of a disaster. Another approach is based on observing the physical and behavioral relations between environment quality indicators and the measurable results of human activity, such as the diminishing availability of resources, or social issues, such as the impact on human health (Giordano, 2006).

Analyses and evaluations of ecological projects discussed in the literature, as well as the methods of evaluation of profit and loss, include (Matczak, 2000) indirect methods (based on an analysis of market reaction) along with direct methods (based on surveys and questionnaires).

The proposals from the first group include the following solutions (Jezowski, 2002; Giordano, 2006):

- Cost-Benefit Analysis (CBA). If a project costs less than its expected monetary profits, it is a beneficial investment.
- Cost Effectiveness Analysis (CEA), which only shows costs in cash values. This method is used in a situation when the benefits are impossible to express in monetary units. This may include for example achieving a certain ecological standard. In that case, the evaluation would only include the costs of the technical investment, which would, under certain circumstances, reduce the level of pollution to that set by the adopted standards.
- Risk Benefit Analysis (RBA) used in situations when taking some action (or refraining from it) carries with it some risk. For instance, refraining from building a water treatment plant means that the available water will be of poor quality and may become a risk factor in the development of diseases among the people who drink it.
- Multi-Criteria Analysis (MCA) is used in reference to investments which include conflicting priorities. Its goal is to achieve the 'maximum consensus'.
- Hedonic Prices Method (HPM) refers to elements of the environment, whose prices are not directly shaped by the market, and whose evaluation is made based on substitute goods. For instance, the price of recreational places around a lake allows for an evaluation of water quality in that lake (which is an element of the environment).
- One variety of this method is the Hedonic Wages Method (HWM), which takes into account the impact of environmental factors on wages in the region.
- Human Resources Method (HMR), which indirectly estimates the value of human life and health, by determining the costs of medical treatment and also the losses associated with environmental conditions.
- Restoration Method (RM) refers to evaluating the costs associated with restoring the lost values of the environment.
- Substitution Method (SM) is used in situations where restoring the lost value of the environment is difficult or impossible. The cost of a substitute for what has been lost is then evaluated.

Such propositions are associated with the interesting idea of 'environmental accounts' (Zylicz, 1997), realized in Europe e.g. in Sweden. It takes into account the loss of various functions performed by the environment and attempts to assess how much restoring the destroyed areas would cost. In order to answer the question, an environmental report is first made, which shows in physical quantities (e.g. in tonne) the level of pollution emissions into the environment and the production of waste. After summarizing the data, using fixed fees regarding the impact these actions have on the environment, the monetary values are calculated (Zylicz, 1997).

Another form of environmental account is the idea of an ecological debt, which is the balance of two values:

- Loss caused by environmental degradation.
- Costs of restoring the destroyed state.

Economists state that if the loss is big, while the cost of restoration is low, we should (as a present to the next generation) pay this smaller amount in order to restore the environment to its previous quality. If the loss is small, however, and the cost of restoration is high, it would be better to skip the restoration option. What has been lost is, in this case, the lower amount (Zylicz, 1997).

Another methodological approach includes studying consumer preferences using surveys and questionnaires. This is called the Contingent Valuation Method (CVM). Two principles find application here (Jezowski, 2002; Giordano, 2006; Zylicz, 1993):

- Willingness to Pay (WTP), which is a question of the maximum amount a person would be willing to pay for receiving (or protecting) a given environmental good.
- Willingness to Accept (WTA), which refers to the minimum amount that a person will accept to refrain from using a particular good, thus stopping its degradation.

Surveys can also help to evaluate places that are attractive in terms of tourism, by analyzing demand for the given locations' values. It is called the Travel Costs Method (TCM), which refers to the distance traveled by tourists in order to reach an attractive destination and the costs which they bear (Jezowski, 2002).

The weakness of these methods is that answering questions is one thing, and actually spending certain sums of money is entirely different. Therefore, the best solution in evaluating environmental value is to choose more than one method.

Paradoxically, even the best environmental evaluation and its inclusion in budget bills, may turn out to be insufficient. A major problem may arise where the free market meets the policies implemented by individual countries (or groups of countries). If one country includes environmental costs in the prices of its products, and another does not, the first one may suffer significant financial loss. That is because most people would choose cheaper products. Let Levi Strauss and Co. serve as an example. The company had, at its own initiative, prepared a widespread strategy for environmental protection, also including social issues. The competition forced them to break some of these provisions, such as that associated with not hiring cheap labor in poor countries (Barbier, 1995).

The prices can be further lowered by reducing wages or increasing the number of working hours. This is a unique combination of economic, environmental and social conditions.

China is often pointed to in the context of the above. Currently this country is conquering global markets. Let us focus on the consequences of an insufficient consideration of the environment in the economic calculations in this country. For example, when Western companies won tender offers for the construction of coal power plants, they were forced by Chinese authorities to lower their environmental standards, which would make the investment cheaper (Bialowas, 2005; Esty & Gentry, 1997; Strandberg, 1992).

A more significant problem is the excessive depletion of water in China— both surface and groundwater. In terms of ecology, it should be pointed out that, although the Chinese constitute 20% of the global population, they only have access to 7% of the world's water supply. Despite these negative conditions, scientists estimate that factories in modernizing China still use from 3 to 10 times more water than similar plants in developed countries. Agriculture also uses too much water there. This is a consequence of a situation long past (it took place just after World War II), when one of the priorities for the government was achieving self-sufficiency in grain production. Enormous plantations were created, and the crops were abundantly irrigated.

In the municipalities water management in China is also unsustainable. For instance, in the very rapidly developing city of Shijiazhuang, characterized by excellent financial results, and inhabited by 2 million people, two-thirds of the groundwater resources have been exhausted (Yardley, 2007). The high quality of the life of the residents of such cities as Shijiazhuang may soon suffer a sudden breakdown. After all, no human settlement can function without a water supply, least of all a two million person agglomeration. This does not, however, change the Chinese pricing policy towards water. The authorities are aware of the problem, but the remedial funds are mainly concentrated on building canals that enable the flow of water from regions with a better water balance to those most threatened. The system is to be completed by 2050, and is estimated to cost around US $ 62 billion. Without an actual change in the approach to environmental calculations, this investment will merely postpone the problem instead of providing an actual solution (Yardley, 2007).

The issues presented above are created by our treating the environment as an 'external' factor. The scale of the problem is shown by discounting. This means determining the value of certain capital in the future, based on present-day prices (Zylicz, 1997). In the case of environmental capital and human activity, discounting shows that although actions for environmental protection are initially characterized by high costs and low benefits, in the long run this situation is reversed.

For instance, if we assume a discount rate of 10%, then the value of the assumed costs today of environmental protection at the level of US $ 1 million spent in a hundred years time would be just US $ 72 (Page, 1997). In other words, this is how much needs to be spent today in order to prevent much larger expenses in the future. In reality, these differences are even more significant, as the costs of environmental protection already amount to billions of dollars. However, should we fail to bear the cost to a sufficient degree, we may face a breakdown (both economic and ecological) in the future, when the necessary amount exceeds our ability to pay.

3.2 *Economic instruments for protection of the environment*

Such ideas as discounting, in order to truly protect the environment, need to be based on certain economic instruments. The budgets of individual states are usually based on three pillars, which are (Zylicz, 1997):

- Profit tax paid for profit achieved by businesses.
- Turnover tax paid by businesses.
- Payroll tax.

Taxes are also an important instrument, in environmental protection, but there are other opportunities as well. It is worthwhile to recall the international classification of such instruments accepted by the OECD (2010):

- Fees and taxes for polluting the environment and issuing permits to use its resources (like extraction fees for groundwater). These fees generally refer to measurable spot sources of pollution. However, dispersed sources are a problem as well, and the way these are evaluated is an important task for environmental economics (Andreasson-Gren et al., 1992).

Moreover, this group also includes product fees. These are introduced with regard to pieces of a product, their weight, or packaging (e.g. plastic bags, or non-returnable bottles and cans) and also fuels.

- Subsidies for environmentally friendly projects, crediting green investments, and tax credits for investments in the environment. We should also call attention to 'eco-conversion', which is replacing (to some extent) the foreign debt of certain countries with green investments.
- Subsides are a specific economic instrument. In the context of sustainable development, they should encourage efforts for protection of the environment, though this is not always the case. There are a number of industries, whose functioning is based largely on direct or indirect subsidies. Furthermore, many development programs assume subsidizing solutions that, although masked as general development goals, are essentially very bad for the environment ('bad subsides'). These are often subsidies for the heavy industry and energetics, which stimulate excessive consumption of resources and fossil fuels, as well as those favoring road transport over public transport, especially railways (Kozlowski, 2005). Bad pricing policy may also refer to agriculture. For instance, profits from subsidies for cane growers in Florida practically allowed them go unpunished for dumping huge amounts of sewage into the waters of the Everglades (Futrell, 2004).
- Creating deposit refund systems, such as introducing deposit fees for car batteries. If people return their old batteries when buying new ones, the fee is not charged. If, however, the old battery is not returned, the fee is required. Moreover, deposits may apply to tires, or plastic bottles. The goal is to prevent their illegal storage or dumping.
- Market creation, like marketable permits, enables the sale of permits to pollute in the area. A limit is set for the level of pollution, leaving the issue of which company will buy what part of the share to the free market.
- Enforcement incentives, stimuli for enforcing the law. These include any kind of non-compliance fees (e.g. for emissions of pollution higher than that allowed), or performance bonds (e.g. to raise funds for future restoration of the environment).

In practice, the most frequently used methods are various fees and taxes, which often do not include external costs, neglect many material turnovers, and do not encourage reducing material and energy consumption (Stodulski, 2004). It is, therefore, important to introduce all the ecological instruments from the groups developed by the OECD.

In this context, we might recall the concept of the 'shift tax' (Hammond et al., 2001), which proposed reducing (or even eliminating) taxes associated with goods that are important for the national economy, and increasing the taxes for products that are considered less important or exclusive. It is therefore a tax that is charged not on the level of pollution of the environment, but on the consumption level. An important factor is that in the rich countries of the OECD, consumption exceeds 70% of GDP, most of which is in the form of durable items, such as houses, cars, or electronic hardware (Albrecht, 2006). Since many companies produce both 'sustainable' and 'unsustainable' goods, charging increased taxes on the latter should be a good stimulus to develop the former. The problem seems to be defining a list of such goods so as to not raise any doubts, on the other hand preparing such a list seems realistic for some goods at least.

132

Will these solutions be enough to achieve the variety of changes needed in the era of sustainable development? Will we create the 'green' GDP?

Even at the level of the EU there are serious doubts. The high level of integration, both political and economic, is surely a success. Historically speaking, such processes usually took much longer. Consider the case of the United States. Political integration took place after the Civil War (1861–1865), but economic integration was not possible for another 100 years (Domanski, 2004). In the case of the European Community, we note that it was established with expected economic benefits in mind first. Other goals, such as environmental protection were not as clearly defined at the time. Some commentators even argue that developing a common market took place at the expense of the proper promotion of environmental protection and sustainable development (Pawłowski, 2001).

An example can be found in the form of one of the key issues, which is the need to change present models of production and consumption. It is enough to mention that in case of one of the EU's richest countries, Germany, the total pressure on the environment from its 80 million citizens, is, at the present level of consumption, higher than that generated by 900 million Hindus (Pawłowski, 2001). Moreover, the reports by the European Environment Agency show that the main production and consumption tendencies in the EU (in Germany as well) have not changed and there is no indication that they would change any time soon (Pawłowski, 2001). It is, therefore, necessary to introduce fundamental changes to the traditional approach to business. The idea of 'responsible business' is a good example (Laszlo, 2005).

3.3 *Responsible business and environmental systems of management*

The 'responsible business' concept (Goodwin, 2001) is supported by such organizations as the Coalition for Environmentally Responsible Economies (CERES), the International Chamber of Commerce (ICC), the Business Charter for Sustainable Development, and the International Labor Organization (ILO). Its basic principles are presented in Table 33.

These principles go far beyond purely economic dimensions, referring not only to environmental issues, but also to social, and even ethical issues. It is a good approach, as economics in itself can only show what is economically rational, it cannot show what is good in terms of human development (Dobrzanski, 2000).

The principles mentioned above are expressed in more detail within a specific accepted system of management. Diagnostic tools are in use here, such as audits, i.e. assessments of the impact on the environment of a given unit and also suggestions regarding possible improvements (Vanclay & Bronstein, 1995; Ortolano & Shepherd, 1995; Boothroyd, 1995; Leistriztz, 1995; Buckley, 1995; Strahl, 1997). The formal aspect is overseen by the International Association for Impact Assessment (IAIA), which was established in 1980 and involves more than 1600 members, the representatives of 120 countries (IAIA, 2011; Hamm, 1995).

An important group of audits are those carried out on individual products that enter the market. The concept of Life Cycle Assessment (LCA) is particularly useful (Zbicinski et al., 2007; Poltorak et al., 2006). It includes the 'cradle-to-grave' approach (Strahl, 1997) and analyzes three major problems:

- Impact of production processes on the environment.
- Impact of products during their use on the environment.

Table 33. Basic principles of responsible business (Rok, 2004).

No.	Principle
1.	Achieving sustainable profit, while shaping reasonable relations with all stakeholders*.
2.	Using such instruments, which enable building a dialog with the stakeholders in order to improve the company's development strategy.
3.	Adopting a philosophy of business activity based on creating sustainable, clear relations with all interested parties (workforce, clients, suppliers, local communities, and even the competition).
4.	Building a strategy of competitive advantage in the market, based on ensuring sustainable value for both shareholders and other stakeholders.
5.	Providing services and products in such a way as to not degrade the environment, both natural and social.
6.	Conducting business in such a way as to include ethical values, adherence to the law, and respect for employees, society and the natural environment.
7.	Fair fulfillment of obligations.
8.	Conducting business in accordance with society's expectations, which are of an ethical character as well as legal, financial, and civic.
9.	Creating and implementing a strategy of social engagement, exceeding legal obligations, for the good of all citizens in accordance with commonly accepted ethical standards.
10.	Conducting business so that its effects are consistent with social expectations and values.
11.	Contributing to sustainable development through cooperation with the workforce, as well as the local and global communities, in order to improve the quality of life of all citizens.
12.	A sense of responsibility for the consumer, the investor, the society, the natural environment, and for economic success.
13.	Voluntarily including social, ethical, and ecological aspects in business and in dealings with stakeholders.
14.	Using transparent business practices, based on respect for employees, society, and the environment.

*A stakeholder is a person or group interested in the operation of a company and taking various risks related to its activity. It is a wider group than the shareholders. Shareholders are chiefly interested in the company's economic profit, whereas stakeholders include employees, clients, suppliers, or even the local community (Forum of Corporate Governance, 2007).

- What happens to the product after it has been used (whether it goes to a landfill or is partly reused).

However, the most information comes from full systems of environmental management. They are part of the general management system and include the organizational structure, planning, responsibility, rules of conduct, procedures, processes, and means necessary to develop, implement, realize, overview, and maintain environmental policies (Borys, 2002), which result in reducing any negative impact on the environment. These types of solutions include (Weis & Bentlage, 2006):

134

- Good Manufacturing Practice (GMP), regarding pharmaceutical products and food.
- Hazard Analysis and Critical Control Process (HACCP), which is concerned with providing security during the production, distribution, and storage of products. It puts an emphasis on hygiene as well.
- Occupational Health and Safety Advisory Service, OHSAS 18001 associated with health and safety rules.
- British Standard BS 8800, compatible with OHSAS 18001.

In Europe, the solutions based on ISO standards as well as the EMAS system are the most popular (Kramer et al., 2004).

The International Organization for Standardization (ISO) is a non-government agency established in 1947 (ISO, 2010). ISO's environmental standards belong to the 14001 series (among others), which was initially introduced in 1996 and modified in 2004 (Klemmensen et al., 2007). A list of exemplary standards is shown in Table 34.

Quality standards (ISO 9001) are also important, as well as ones addressing the level of Corporate Social Responsibility (ISO 26000) introduced in 2005 (Weis & Bentlage, 2006).

Another significant model of environmental management is the European Eco-Management and Audit Scheme (EMAS). The first version of EMAS was introduced in 1993 (Regulation no. 1863/93) and EMAS II in 2001 (Regulation no. 761/2001), modified in 2010 (Regulation no. 1221/2009). Receiving an ISO certificate within EMAS exempts organizations from some obligations which overlap with the EMAS requirements.

An integral part of the EMAS system is the preparation of environmental reports, which are the bases for the formulation of particular action programs. The major questions at the level of such report are (Borys & Rogala, 2002):

- What type of operation does the unit perform?
- Can the efficiency of the tasks carried out be improved (e.g. in terms of resource consumption)?
- What is the plant's impact on the environment?
- What goals does the unit realize in terms of environmental protection?
- What are the effects of realizing these goals?
- How is environmental management carried out?

Table 34. Some ISO standards regarding chosen environmental aspects (Author's own work).

Issue	Examples of documents
Environmental management systems	ISO 14001, ISO 14004, ISO 14061
Environmental audits	ISO 14010, ISO 14011, ISO 14012, ISO 14015, ISO 19011
Environmental performance assessment	ISO 14031, ISO 14032, ISO 14063, ISO 14050
Other environmental issues	ISO 14062, ISO 14050

This last aspect includes three subject scopes (Borys & Rogala, 2002):

- What tools are in use in the unit?
  There are diagnostic tools (e.g. LCA, audits, and environmental reports) and implementation tools (e.g. eco-labeling, cleaner production).
- What do they refer to?
  There are two options—reference to technological processes or to products.
- Are the actions taken formalized or not?
  In the case of the former, are they based on standards such as ISO?

In the EU, the first step is often to seek an ISO certification and then carry out a full audit implementing the EMAS system.

Comparing the ISO and EMAS systems, we can point out several important differences (Weis & Bentlage, 2006; Strahl, 1997):

- ISO standards refer to all countries in the world, whereas EMAS is an internal solution adopted within the EU.
- ISO standards are managed by a non-governmental organization, whereas EMAS is a system implemented at the governmental level of individual countries.
- EMAS generally refers to production plants, while ISO refers to all interested organizational units.
- In both cases, meeting legal standards is required. However, in the case of ISO it is only environmental law, whereas EMAS requires compliance with the whole legal system.
- The EMAS standard stresses the necessity to make the entire implemented system public (publishing environmental reports). ISO provides only general conditions. Hence the pressure is on to not use specialist language in the case of EMAS, in order to make the message clear for all recipients.
- EMAS draws attention to the need to engage the employees of a given unit in the process, as they usually know the problems occurring there.
- Introduction of an environmental management system (both ISO and EMAS) is also possible in plants characterized by high levels of pollutant emissions. Although the organization's way of functioning (especially in the economic aspect) is improved, this does not always correspond to significant reductions in the negative impacts on the environment.
- The ISO trademark is widely recognized, but EMAS is still largely unknown, which is a problem considering the multitude of popular eco-labels.

The last point is particularly important, as implementing both solutions, especially EMAS, is expensive. If they are spending money, manufacturers expect that implementing an environmental management system will give them benefits, including increased company prestige (Goodwin, 2001). This is not always the case, especially from a short-term perspective. Therefore, other economic instruments, like tax relief, are important. These can be granted by governments to companies that introduce EMAS, but in some EU countries there are no such additional stimuli.

It should also be noted that an absence of ISO or EMAS certification does not mean the plant is badly managed; it means only that the organization is not able to meet at least one of the requirements of these systems, or has insufficient funds for implementation (Williams & Dair, 2007). What is also important, when the system is not implemented and the plant's impacts on the environment are undefined or their balance is negative, is that banks and insurance companies may withhold a decision on providing credit to the company, which might, in turn, affect its development (Strahl, 1997).

Complex environmental management systems are not the only economic aspect affecting the sales of given company products. It is also important they be given the 'ecological labels' that usually refer to particular products. A selection of such labels is presented in Table 35.

Another important system is the labeling of light bulbs, fluorescent lamps, and electronic hardware according to their energy consumption. The EU standards distinguish classes from A to G (Weis & Bentlage, 2006). The higher the class, the more expensive the device, but the cheaper it is to use because of its lower energy consumption.

The Material Input per Service (MIPS) index is also an interesting concept for an eco-label (Zbicinski & Stavenuiter, 2006). This concept was formulated at the Wuppertal Institute for Climate, Environment and Energy (Karlsson, 1997; Hinterberger & Schmidt-Bleek, 1999). It shows how much and what kind of materials must be used in order to manufacture a particular product. It refers to both consumer goods and industrial materials. The higher the index, the higher is the consumption of materials to make the product. The calculation also includes the cost of acquiring resources, their processing, and transport. Within a group of consumer products, it can easily show goods such as yoghurts, which can be made practically anywhere, but which, at the will of large consortia, sometimes 'travel' over vast distances, even throughout Europe. Also, within the industrial group, the index shows that the production of 1 metric tonne of

Table 35.   Chosen eco-labels (Zbicinski & Stavenuiter, 2006).

Name	Country	Year of introduction
Blue Angel	Germany	1978
Environmental Choice	Canada	1988
Environmental Choice	New Zealand	1990
Good Environmental Choice	Sweden	1990
Ecomark	Japan	1989
Ecomark	India	1991
Ecomark	Korea	1992
White Swan	Nordic Countries	1989
Green Seal	USA	1989
Green Label	Singapore	1992
EU Flower	European Union	1992
Stichting Milieukeur	Netherlands	1992
NF-Environment	France	1992
CE (Communauté Européenne) Product labeling ensuring the product is safe to use	European Union	2001
Polish EKO-sign	Poland	2005*

*In this year, the label was adapted to criteria in force within the EU.

steel takes 53 tonne of materials, including 7 tonne of resources and 44.6 tonne of water (Weizsäcker et al., 1999). It is also worth noting that although MIPS has a quantitative character (not a qualitative one since it does not include such issues as the toxicity of the materials and products).

MIPS has not gone beyond the experimental stage so far, but it still shows that consumers may soon get brand new tools allowing them to quickly assess a particular product's impact on the environment (Zbicinski & Stavenuiter, 2006).

However, the economic aspect of sustainable development goes far beyond determining the balance of costs at the product level (if only regarding eco-labels), or the system of environmental management at the level of a particular plant. It is also the issue of financial security, which would enable the realization of numerous agreements and conventions at the national level, or adopted internationally, usually within the UN.

3.4   Financial security for introducing sustainable development

The most famous global framework program of implementing sustainable development is "Agenda 21". It was passed in Rio de Janeiro in 1992 and assumed the task of raising funds for the realization of the goals set—some US $ 600 billion a year (Baltscheffsky, 1992). A large part of the money was to be paid by the rich countries by collecting 0.7% of their annual national income. Objections were raised by some countries, including the USA, whose later contribution to "Agenda 21" was no more than 0.21% of their income (paradoxically, this was not a small amount given, because of the country's high total income). Only the Nordic countries and Netherlands fulfilled their obligations (Baltscheffsky, 1992; Leva, 2004).

Another important UN initiative was the "Millennium Development Goals Report", approved in New York in 2000. This was a sort of expansion of "Agenda 21" and it has not been fully implemented for the same reason—the inability to raise the necessary funds (Baltscheffsky, 1992; Leva, 2004).

Of course, there are new financial opportunities available, reaching further than the previously accepted OECD instruments listed in Table 36.

The question is, will there be enough political will to introduce them? Consider, for example, just one of these proposals—the Tobin tax (levying a relatively low fee on international financial transactions). Assuming the fee would be just 0.1%, this would yield around US $ 300 billion a year (Kozlowski, 2005). This would provide half the amount required to cover the implementation of "Agenda 21". Also it must be remembered that about US $ 1.2 trillion a year are spent globally on the military (Marzec, 2008) and another US $ 1 trillion goes as subsidies for actions that are bad for the environment, including subsidizing ineffective energy management and developing industries that are material-consuming and pollute the environment (Andersson, 1997).

The disproportion between the relatively low amounts spent on programs supporting sustainable development and the huge amounts spent on completely different goals, contributes not only to the increasing poverty in the countries of the South, but also to their increasing debts.

This is largely the result of the enormous capital surplus that appeared in the market in the early 1970s, related to a sudden increase in the price of crude oil (Sachs, 1993). The countries exploiting this resource started investing their additional incomes in Western banks, which in turn were willing to lend money

Table 36. Proposals for new economic instruments for funding the implementation of sustainable development (Kozlowski, 2005).

No.	Proposed instrument
1.	Tobin tax—applied to international financial transactions, e.g. bond trading.
2.	Tax on international trade.
3.	Taxing certain goods, e.g. fuels that are subjects of international trade.
4.	Taxing the international arms trade.
5.	Tax on postal and telecommunications services.
6.	Tax on national taxes (if only some of them, e.g. the tax on luxury goods).
7.	Fees for keeping satellites in orbit.
8.	Fees for exploiting minerals in international waters.
9.	Fees for fishing in international waters.
10.	Fees for carrying out research and exploiting resources in Antarctica.
11.	Tax on international plane trips (or tax on aviation fuel).
12.	Tax on international ship cruises.
13.	Tax on international trade in emission permits.
14.	Fees for using radio waves (telecom, radio, TV).
15.	Voluntary local tax paid to the global fund.
16.	Allocation of Special Drawing Rights for poor countries, for keeping the peace, or for other global public good.
17.	International lotteries.
18.	Selling part of the gold reserves possessed by the International Monetary Fund.

* Special Drawing Rights (SDR)—a conventional cashless monetary unit of the International Monetary Fund, for financial settlements with central banks. Their value is set based on a basket of currencies comprising the American dollar (44%), Euro (34%), British pound (11%), and Japanese yen (11%). SDR units have so far been emitted twice (in 1970–1972 and 1979–1981). A third allocation has been discussed since 1997, but it is, however, blocked by the USA.

to the developing countries (more precisely, their governments, which were in most cases dictatorships). The capital distributed in such a way amounted to about US $ 540 billion (Fair Trade, 2007). During the 1980s, the conditions for debt repayment became worse. Interest rates exceeded 20%, and resource prices fell by 30%. Many countries lost their financial liquidity and were not able to pay both their basic loans and the interest, which rapidly increased the scale of the debt. It also turned out that, although these countries eventually paid more than they had borrowed, their debts were still growing. It is worthwhile to quote a popular Brazilian saying: 'Brazil borrowed US $ 100 billion, paid US $ 100 billion and still owes US $ 100 billion' (Wise, 2001).

At the beginning of the 21st century, African countries were approximately US $ 230 billion in debt and were spending US $ 39 billion on payments. This expenditure is bigger than their outgoings for other developmental goals such as education or healthcare (Ryden, 1997). In other words, it is impossible to reduce impoverished areas in these countries without reducing their debt.

One of the more important assistance programs is the Heavily Indebted Poor Country Initiative (HIPCI). This program, supported by the International Monetary Fund and the World Bank, was established in 1996. It proposed reducing the debts of countries of the HIPCI to between 20 and 25% of these countries, actual export incomes (IMF, 2010). In 2005, the program was expanded by the Multilateral Debt Relief Initiative (MDRI), which enabled countries that were previously subject to HIPCI to further reduce all their debts to the International Monetary Fund and the World Bank. The project also involved the International Development Association (IDA) and the African Development Fund (AFC), both associated with the World Bank. The reduction totaled US $ 56 billion and covered the 32 poorest countries, 26 of which are in Africa (IMF, 2010).

Independently of the debt reductions, support should also be given to specific strategies, such as the effective use of the ability to work, which is an important 'resource' of the poor (Heredia, 1996; Lothigius, 1995).

It is also important to support the poorest countries outside official international aid. A good example of such help is the famous "Live Aid" charity concert of July 13, 1985. The idea was introduced by Bob Geldof (leader of the rock band The Boomtown Rats) and Midge Ure (co-founder of the new-romantic band Ultravox). The two main concerts took place at the Wembley Stadium, London (72,000 viewers) and the JFK Stadium, Philadelphia (90,000 viewers). Estimates show that TV broadcasts allowed about 1.5 billion people in 100 countries to watch the show.

The concert in London starred e.g. Status Quo, The Boomtown Rats, Ultravox, Spandau Ballet, Sade, Phil Collins, Bryan Ferry, Paul Young, Allison Moyet, U2, Dire Straits, Queen, The Who, Elton John, Paul McCartney, and in the finale, Band Aid with the charity song "Do They Know it's Christmas?". In Philadelphia, the concert involved e.g. Joan Baez, B.B. King, Black Sabbath, Phil Collins, Judas Priest, Bryan Adams, The Beach Boys, David Bowie, Mick Jagger, Simple Minds, The Pretenders, Santana, Madonna, Tom Petty, Led Zeppelin, Crosby, Stills, Nash and Young, Duran Duran, Tina Turner, and, in the finale, USA for Africa with the charity song "We are the World".

"Live Aid" raised almost US $ 300 million, which was given to charity. This amount was merely a drop in the bucket. Nevertheless, it drew the attention of the global society to the problems of Africa.

"Live Aid" concerts had continuation on July 2, 2005. The event had been deliberately organized just before the G8 summit, hence the name "Live 8". The main objective was to draw further attention to the worsening situation in Africa. This time no money was collected; the organizers wanted viewers to sign a petition to leaders of the G8 countries that they should double the level of aid to poor nations, and 'make poverty history'. The concerts took place in London, Philadelphia, Berlin, Rome, Paris, Barrie (near Toronto) and Edinburgh (this last on July 6, 2005). The concerts involved e.g. U2, Elton John, Pink Floyd (with Roger Waters), The Who, Madonna, Paul McCartney, Duran Duran, Crosby, Stills and Nash, Bon Jovi, Stevie Wonder, Bryan Adams, and Deep Purple.

It is not only such large-scale events that may be helpful. The financial institutions that can individually support small-scale environmental and infrastructural projects in practically every region of the world are also of great importance. Of course, they are not always successful.

An example can be found in how the issue of access to clean water was to be solved in many African countries. It is a major issue, since it concerns about

1 billion people. It is estimated that infected water is the cause of 80% of the diseases in developing countries and kills about three million people every year; this is more than the number of people who die due to HIV infections (Water Safe ..., 2004). In order to help Africans, water-treatment facilities were built in vulnerable regions. These significantly improved water quality. Later on, however, these were privatized. The prices set by the new owners were too high for the residents. In practice, just as they had no access to clean water in the first place, so they have none still.

Such situation happened e.g. in South Africa just a decade after the political breakthrough from the year 1994, when Nelson Mandela become a president of this country (Klein, 2007).

It supposed to happen also in Bolivia. In the city of Cochabamba privatization of water services was done in the end of 1999, and it lead to fast price increases. This was the reason for the formation of La Coordinadora de Defensa del Agua y de la Vida (The Coalition in Defense of Water and Life) in January 2000. After enormous demonstrations people finally won on April 10, 2000 (VanOverbeke, 2004). Unfortunately, in 2005, still half of the people in this city remained without water, and the rest with only intermittent service (Ferero, 2005). In such cases cheaper technologies may help, not so much in treating polluted surface waters, but in storing rainwater (Sharing Sustainable Solutions, 2011). This, however, does not fully end the problem.

Many more benefits can be drawn from other programs, such as "The Equator Principles". It is an initiative of the world's 20 largest financial institutions and it directed at achieving minimum environmental and social standards. As part of this project, large grants are prepared (exceeding US $ 10 million) for developing countries. They are given based on three criteria that are in turn based on 10 principles (Lawrence & Thomas, 2004).

These criteria address the expected scope of the changes introduced, their possible effect on the environment, and their effect on the local community. The principles are presented in Table 37.

"The Equator Principles" make a full system (general principles are enforced with a whole system of indicators) clearly defining the criteria for financing projects and providing a control carried out by an independent institution. Although some grants are being criticized by experts (Lawrence & Thomas, 2004), "The Equator Principles" project is generally marked as positive. The diversity of the actions taken should be highlighted, as well as their close relation to sustainable development. They cover searching for raw materials, new production methods, renewable energy, agriculture, and forestry (Lawrence & Thomas, 2004). A stable financial background is also important, as it guarantees the completion of each grant.

It is worth comparing the proposals presented with the European perspective.

For many years the region was divided between the free market-oriented West and the communist East. The changes, started in Poland in 1989, were definitely a breakthrough point (Rosati & Bochniarz, 1991; Zylicz, 1991). They were associated with the implementation of the so-called 'Balcerowicz plan', which was based on six pillars (Sachs, 1994):

- Macroeconomic stabilization.
- Liberalization; dropping the idea of central planning.

Table 37.   The equator principles (Equator Principles, 2006).

No.	Short description
Principle 1.	Project review, determining its possible impact on the natural and social environment, including possible hazards. These tasks are carried out by the creditor.
Principle 2.	Audit of the project's impact on the social and natural environment, carried out by the borrower.
Principle 3.	Checking whether the project meets social and environmental standards set by the creditor.
Principle 4.	The borrower prepares an action plan and the project management system.
Principle 5.	Consultations with the communities, to which the project refers.
Principle 6.	An appeal mechanism is created that would allow citizens and local communities to lodge objections to the project.
Principle 7.	Independent project review, carried out by an institution other than the borrower.
Principle 8.	Preparing legal documents compatible with the project as well as with the existing legal system.
Principle 9.	Monitoring the realization of the project independently of the borrower.
Principle 10.	The creditor publishes, at least once a year, a report on its activities concerning the projects funded.

- Price control.
- Privatization.
- Getting foreign financial help.
- Creating a network of social security, including unemployment benefits.

The program was close to the liberal "Washington Consensus", which, despite equally ambitious goals, has led to an economic collapse in the countries of Latin America[10] (Klein, 2007).

In Poland the realization of the program went more ambiguously. Many commentators noted that emphasis was put on the first five pillars. The sixth was only partly introduced (Barber, 1995) due to the massive social costs that would have to be faced (the failure of many large industrial plants, and growing unemployment). However, economic (as well as political) success was achieved. It was these changes that preceded the later changes in other Eastern countries that enabled the expansion of the EU.

In the case of the integration of Eastern and Western European countries let us note that, initially, the major obstacles were considered to be political issues. It quickly turned out that the differences in the general level of development were more important. These included infrastructure issues which were largely

---

[10] The Consensus was an initiative of the World Bank, the International Monetary Fund, World Trade Organization and the US Treasury. The reaction to this spectacular failure was the creation of the Multilateral Debt Relief Initiative Fund (mentioned earlier in this chapter) by the very same financial institutions.

determined by a lower economic power. Therefore, special institutions and assistance programs were established that sought to reduce these disparities. Three were of special importance (Andersson, 1997).

- The Global Environmental Facility (GEF), established in 1991, which is a joint enterprise of the World Bank and two UN agendas—the United Nations Environmental Programme (UNEP) and the United Nations Development Programme (UNDP). It is designed to solve major environmental challenges, such as preventing the loss of biodiversity, protection of the ozone layer, and reduction of greenhouse gases emissions.
- The European Bank for Reconstruction and Development (EBRD), established in 1991. Its goal is to support action in terms of the environmental protection infrastructure.
- EU programs for candidate countries:
    - PHARE, established in 1989. It originally only referred to Poland and Hungary (the acronym stands for Poland and Hungary Assistance for Restructuring their Economies), though, following the transformations taking place in East Europe, more countries benefited from it.
    - It is worth remembering the special version of this program, designed for countries of the former USSR—Technical Aid to the Commonwealth of Independent States (TACIS), which has functioned since 1991.
    - Instrument for Structural Policies for Pre-Accession (ISPA), launched in 2000.
    - Special Accession Program for Agriculture and Rural Development (SAPARD), established in 1999.

The enlargement of the EU meant that new member countries gained access to separate structural and special funds and helped determine the development of the Community. The Cohesion Fund, which is the EU's structural tool, plays an important role here. It was established in 1994 by the "Treaty of Maastricht", which is the same document that enabled the creation of the EU (Agenda 21, The First Five …, 1997) The Cohesion Fund refers to projects at the national level. Smaller programs are realized within the European Regional Development Fund (ERDF, 2010).

The examples of the economic tools presented above, as well as other funds or programs, make modern economies more and more 'green'. It is also important to introduce principles, such as the mentioned "Principles of responsible business", which refer directly to social issues. After all, society is currently serving economics, while it should be the other way around (Hull, 2007). It is an issue of re-evaluating the presently binding economic hierarchy of values. This leads us to the special significance of such concepts as the personological economy, developed in Poland by Franciszek Piontek. This is defined as a science on the laws of management, in which humans play the role of subjects and creators of the process. It is an analysis of economic processes, which is taking into account the subjective and creative functions of humans (Piontek & Piontek, 2007). This is a transition from *homo economicus* to *homo sustinenes* (Fiodor, 2005).

In practice, much will depend not only on the available economic means, but also on the constantly changing technical capabilities of human beings, the legal conditions, as well as the political will to act. This will be the subject of analysis in the next part of the book.

CHAPTER 5

Level III of sustainable development: Technical, legal and political considerations

All strategies of development depend on political will and legal system. From the practical point of view we must also include available technical solutions. This is the third level of sustainable development.

1 TECHNICAL PLANE

1.1 *Technology and environment*

Technology is an area of human activity whose objective is to produce and generate objects, as well as phenomena not found in the natural environment, and to modify nature (Universal Encyclopedia, 1984).

The ever-increasing technical capabilities of the human race are well reflected by statistics and historical data. The 19th century saw as many inventions as all previous generations in total. In the 20th century, the technical development gained new momentum—technology become so powerful, that even the destruction of the whole planet become possible. Will these powers be used not for destruction, but for the good in the era of sustainable development?

The development of science, technology, and the related development of industry was the main factor contributing to the degradation of the natural environment. At the same time, however, the basic environmental protection strategies are implemented at the technical level.

Failure cannot be avoided altogether.

The area of water protection is an example. Since 1950s, new problems would arise almost exactly every ten years: reduction of oxygen content in water, eutrophication, heavy metals pollution, acidification, and contamination of ground waters (Hultman, 1992; Forsberg, 1992). Despite a number of warning signs, the condition of the environment deteriorated steadily, both in the central-planning Eastern European states and in liberal Western countries. The European Union did not solve these environmental problems either. The use and exploitation of the natural environment still leads to a number of negative environmental impacts, both direct and indirect, reversible and irreversible, and of different scale and extent: from minor impacts of local events up to far-reaching global problems.

Local and regional environmental pollution leads to the formation of smog, acid rain, and eutrophication. The first two phenomena fall within the scope of air protection.

Smog is created when the concentration of pollutants in air is high and when there are specific weather conditions (e.g. no wind), as well as climatic and topographic conditions hindering the dispersion of pollutants (a valley rather than an open area).

Two types of smog are distinguished: 'Los Angeles' and 'London' smog):

- 'Los Angeles' smog (first officially recorded in the City of Angels)—also known as photochemical smog, is typical of tropical and subtropical climates and to a large extent related to the increase in road traffic. Main pollutants include: nitrogen oxides, carbon oxide, and hydrocarbons. Solar radiation is the critical factor necessary for the formation of this type of smog. As a result of photodissociation of NO_2 and reaction with atmospheric oxygen, a strong oxidant is created—ozone.
- 'London' smog is also known as acid or sulphuric smoke and usually occurs in moderate climate zones. It is related to high emissions of SO_2, CO, and concentration of soot particles. High humidity is a critical factor. In fog droplets, SO_2 is oxidized to SO_3 and then to the very aggressive sulphuric acid aerosol.

Acid rains are connected with the emissions of sulphur oxides and nitrogen oxides to the air. In humid conditions inside the cloud, H_2SO_4 and HNO_3 are formed and 'washed out' with rain in diluted form (O'Neill, 1993). Acid rains are a threat not only to living organisms (e.g. forest die-back in Central Europe), but also to elements of the human-made infrastructure (e.g. they cause corrosion of railway tracks).

Finally, eutrophication is a major threat in the area of water protection. This phenomenon is related to water bloom, caused mainly by the excessive amounts of N and P compounds (found in fertilizers) washed down from fields. It is typical of the Baltic region, affecting not only the Baltic Sea, but also lakes and rivers (Forsberg, 1992).

In addition to the above local threats, global problems, such as ozone depletion and climate changes, are also the subject of debate.

The ozone hole is a (seasonal or steady) reduction of ozone concentration (one of the greenhouse gases) in the stratosphere (even by 97%). The primary causes of ozone depletion are freons and halons, i.e. compounds not found in the natural environment. They are decomposed in the stratosphere, releasing atomic chlorine and bromine—chemical elements that destroy ozone molecules (one chlorine atom can destroy up to 100,000 ozone molecules). In natural conditions, chlorine would not be found at such high altitude in the atmosphere—it is a highly reactive element and would be oxidized much lower. (Meadows et al., 1995; Harries, 2000).

The greenhouse effect is connected with the concentration of the greenhouse gases (GHG) in the atmosphere. Short-wave solar radiation, that heats up the Earth, penetrates through GHGs. The heat so generated (long-wave radiation) will be to a large extent reflected by the GHG layer back towards the Earth's surface (this phenomenon is known as back-radiation). The mechanism is the same as in a traditional greenhouse—where glass panes have the same effect as greenhouse gases.

Greenhouse gases include (Greenpeace, 1994):

- carbon dioxide (50% share in the greenhouse effect),
- methane (18%),

- freons (14%),
- ozone (12%),
- nitrogen oxides (6%).

We should add to this water vapour—which is also a GHG and its content in the atmosphere increases along with the increase in global temperature (O'Neill, 1993).

50% of the greenhouse effect is attributed to carbon dioxide; therefore, it would be worthwhile to present countries that are responsible for the highest largest CO_2 emissions—see Table 38.

In general, global CO_2 emissions in 2007 reached 27, 245, 758 thousand tonnes, of which 56% was generated by the China, USA and the EU.

We should note at this point that the greenhouse effect is a natural process. It keeps the average temperature on Earth at 14–16°C. Without greenhouse gases, as a simple function of the planet's distance from the Sun, the temperature on our planet would be lower by about 44°C (Universal Encyclopedia, 1984), making it impossible for life on Earth to develop.

However, excessive emissions of GHGs from anthropogenic sources may create a problem. They can increase the warming effect to undesirable levels, melt the polar caps, cause flooding of the vast coastal zones of seas and oceans, or increase the area affected by climate anomalies. Other problems include a decline in agricultural production (due to the increased population of the wintering pests) as well as certain diseases (fungal, bacterial, and viral), and seed germination problems caused by the rising temperature.

The share of anthropogenic gases in the total GHG pool, as well as the impact of other factors on climate changes, are still a matter of discussion (Lindzen, 2010). These other factors include e.g. deforestation, and in particular the destruction of rainforests. It is common knowledge that during the 20th century, as much as 50% of the rainforests were destroyed due to deforestation (Kalinowska, 1992), about 12 million hectares are removed every single year, and about 50 hectares are destroyed every minute (Kostka, 1993). As forests are known for their climate stabilizing function, removal of such a vast portion of their global resources must have a crucial destabilizing impact on the Earth's climate!

In response to the above environmental problems, a number of international agreements were signed to reduce the emissions of harmful compounds,

Table 38. Top pollutants—list of countries by CO_2 emissions in 2007 (CDIAC, 2008).

Rank	Country	Emissions [in thousands of tonnes]	Percentage of global emissions
1.	China (including Taiwan and Hong Kong)	6,538,367.00	22.30
2.	United States of America	5,838,381.00	19.91
3.	European Union	4,177,817.86*	14.04
4.	India	1,612,362.00	5.50
5.	Russia	1,537,357.00	5.24

* of which Germany (highest individual emissions in Europe): 787,936.00 thousand tonnes.

and new technical solutions based on scientific co-operation are promoted in a consistent manner. These solutions included the introduction of new-generation treatment equipment (the 'end-of-pipe' solutions), as well as modifications of the technological process itself. The latter type of solutions offer major savings.

The famous Eiffel Tower in France, constructed in 1889, is one example. Using today's technologies (such as metal alloys), we would have needed only 1/7 of the total amount of steel actually used to construct the tower (Karlsson, 1997).

A number of practical solutions can be employed also today. For instance, technical solutions may be put into practice in large cities to build houses using materials that can be easily demolished in the future, if necessary, and reused in other construction projects (Peski, 1999). These matters are part of the recently developing 'industrial ecology' concept.

1.2 *Industrial ecology and cleaner production*

Robert A. Frosch indicates that industrial ecology is based on an analogy with natural ecosystems. In nature, an ecological system operates through a web of connections in which organisms live and consume each other and each other's waste (Frosch, 1980). The structure of industrial and economic systems is commonly thought to be much different than that of ecological systems. However, supporters of the industrial ecology concept argue that the opposite actually holds- industrial metabolism corresponds to ecological metabolism. Both are thermodynamic systems that exchange energy and raw materials with their environment and with one another (Allenby, 1992; Garner & Keoleian, 1995; Thomas et al., 2003; Ayres, 2004). According to Hardin Tibbs, the aim of industrial ecology is to interpret and adapt an understanding of the natural system and apply it to the design of the man-made system (Tibbs, 1993). We should note that the traditional approach to production processes is linear: production uses raw materials and generates waste. Industrial ecology proposes a closed system where waste is used as a source of energy or substrate for another product (Tibbs, 1993).

With the help of industrial ecology new strategies for the management of raw materials and production processes were formulated. They include (Karlsson, 1997; Wernick et al., 2001; King et al., 2006):

- Reducing the flow (quantity) of raw materials. It boils down to the introduction of material- and energy-saving technologies (the so-called 'dematerialization'). For instance, increased voltage in copper power cables reduces the quantity of copper required while maintaining the full functionality of cables.
- Slowing down of materials flow—better quality of products with fewer defects (higher reliability) and longer useful life.
- Closing of materials flow—use of the recycled materials. It may be based on the introduction of various forms of recycling.
- Substitute the flow—replacing harmful materials with less harmful ones, rare materials with more common ones, or non-renewable materials with renewables (so-called 'trans-materialization').

These strategies are integrated by introducing the 'cleaner production' concept, also referred to as 'eco-efficiency'[11] (Hukkinen, 2001; Clark, 2007; Ropke, 2001; Borys & Bryczkowska, 2003).

Cleaner production refers to three problem areas (Nilsson et al., 2007; Tiberg, 1992):

- Production processes—especially in the context of raw materials and energy savings, elimination of toxic materials, and reduction of emissions of pollutants. Innovations are also important, as is the principle of application of BAT—Best Available Technology (Hellström, 2007).
- Products—reduction of environmental impacts throughout the entire product life cycle.
- Services—selection of services with the lowest environmental impact, or even elimination of certain services.

In general, the goal of cleaner production is to improve quality and efficiency, reduce risks for humans, and achieve cost savings—see Figure 14. The

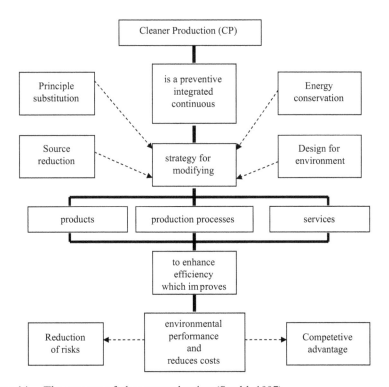

Figure 14. The concept of cleaner production (Strahl, 1997).

[11] The concept of cleaner production was adopted by the UN (UNEP) in 1989, and eco-efficiency was introduced by the WBCSD (World Business Council for Sustainable Development) in 1992 (Klemmensen et al., 2007).

last element deserves special attention: cleaner production is not only more environmentally friendly, but also cheaper, compared to older technologies.

The idea of cleaner production is adapted in a number of areas of science and research.

For instance: the principles of 'green chemistry', sometimes referred to more broadly as 'green engineering', are discussed (Garcia-Serna et al., 2007).

This concept is defined as the design of chemical processes and products that reduce or eliminate the use and generation of hazardous substances (Anastas & Werner, 2005; Nilsson et al., 2007; Paryjczak, 2008).

These ideas are also the cornerstones of environmental engineering (Pawłowski, 2010).

According to an interesting definition, adopted by the Environmental Engineering Committee of the Polish Academy of Sciences, it is a technical science based on the use of engineering methods (Pawłowski, 2007, 2010; Pawłowski et al, 2006, 2007, 2010):

- For preserving, rational shaping and using the external natural environment (e.g. water resources, waste management, air protection, soil protection).
- For preserving and shaping internal environment of buildings and constructions (devices and installations).

In these areas, environmental engineering implements environmentally-friendly initiatives, modifies the technical conditions and technologies to maintain the optimum biological balance of the environment, neutralizes the effects of natural disasters (floods, droughts, or water, air, and soil pollution), and eliminates or reduces the negative environmental impacts of the industrial activities of humans.

The cement industry is one example of the successful introduction of new engineering solutions. It must be noted that from the technological perspective, combustion of energy sources (traditionally in this industry fossil fuels) is the key environmental impact in the production of cement. During the thermal processing of marl and chalk, an important substrate is produced: clinker, but at the same time carbon, sulphur and nitrogen oxides, as well as hydrocarbons and dusts, are emitted to the atmosphere (Pawłowski & Pawłowski, 2004). In the context of cleaner production, owners of cement plants would not only install more efficient treatment filters, but also introduce major technological modifications, for instance to reduce heat consumption. It also means that the amount of fossil fuels burned in cement kilns will be reduced. Major savings were also achieved by using the 'alternative fuels', i.e. waste that cannot be recycled, but can be still used as energy carriers. Alternative fuels include a wide range of waste materials, such as the rubber industry waste, used tyres, timber waste, textiles, waste oils, solvents, and paints, waste generated by the food, paper, and furniture industry, combustible materials from scrapped vehicles, as well as dehydrated sludge. Despite the use of this type of waste, the quality of cement is still high. Environmental benefits are also very significant (Pawłowski & Pawłowski, 2004):

- Reduction of waste deposited in landfills.
- Physical elimination of waste and recovery of energy.
- An option to permanently combine ashes with clinker—it is a nearly waste-free technology.
- Efficient use of fossil fuels.

150

These solutions will play a vital role in the context of the shrinking resources of traditional fuels and the steady increase in their prices.

However, there is still a major obstacle to the effective promotion of cleaner and more efficient technologies: the maximum level of global consumption of raw materials has not been defined. Limits are even more difficult to impose if resources are owned by a number of different entities: countries, corporations, or individual users.

Fisheries may be a good example of such a solution. Limits were introduced when a disastrous decline in fish catch was recorded. The question of exhaustibility of natural resources may be presented in a similar manner. Even the most efficient use of resources will still ultimately lead to their complete consumption. Therefore, the search for various substitutes assumed in the 'cleaner production' concept will gain particular importance in the coming years (Graedel & Klee, 2002). Still, not all resources can be replaced—water being the most prominent example. This aspect must also be considered when formulating the technical strategies of sustainability.

1.3   *Energy issues*

Other major technical challenges will include aspects related to production and excessive consumption of energy, especially electricity. It must be noted at this point that the use of some energy sources is more problematic for the natural environment than the use of others. It was in the 18th century that people started to use coal on a large scale, both in factories and in their homes. In the 19th and throughout the 20th century, oil and gas played a key role, followed by nuclear energy in the second half of the 20th century. In more recent years, we have witnessed the switch to energy production from renewable sources, mainly resulting from the global debate focusing on environmental protection and (later) environmental engineering (Pawłowski, 2010).

On a global scale, coal-based energy production is still the prevailing technology (69% share)—see Figure 15.

The high dependence on technologies based on the combustion of fossil fuels is the underlying cause of many environmental problems of the world today, including water, air, and soil pollution. Note that coal burning is not only a source of dusts and gases emitted to the air, but also of radioactive substances, mainly uranium and thorium (Gabbard, 1993).

Also, coal mining involves the extraction of additional minerals and contaminants, which is impossible to avoid. This is the 'ecological rucksack'— the flow of materials that are not directly used in the product, but are necessary to extract the raw material, or that represent a by-flow that cannot be avoided. Every year, about 60,000 Mtonnes of raw materials are extracted from the lithosphere, of which only 20,000 Mtonnes are raw materials—the rest forms the ecological rucksack Ryden, Mifula, Andersson, 2003; Karlsson, 1997).

In case of hard coal the share of it in the total quantity of materials actually extracted is only 16.7%. But it may be worse. The largest ecological rucksack is extracted during the mining of gold and platinum, which represent only 0.0003% of total quantity of extracted minerals (Ryden et al., 2003; Karlsson, 1997).

The unwanted material is deposited in huge mine spoil-heaps, which are then covered with soils and plants. Still, this waste does not disappear.

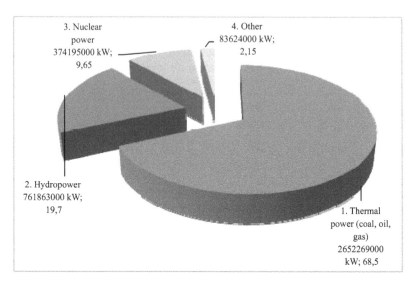

Figure 15.   Total installed electricity capacity by type (International Energy Annual, 2008).

In the context of sustainable development, the discussion on power plants that use fossil fuels should include the problem of the ever-declining sources of energy.

According to recent studies (Pawłowski, 2010), the global resources of coal will be exhausted in 2140, resources of crude oil in 2047, and resources of gas in 2068 (Europe's Energy Portal, 2010)—see Tables 39, 40 and 41. In reality, these time limits may be longer, but the problem remains: the global resources will be exhausted, and it will not take a very long time.

From the perspective of sustainable development, the search for alternative energy sources is therefore the top priority. Technologies based on the reduced use of conventional fuels are also valuable.

This solution refers to the earlier ideas such as the notion of 'negawatt'—i.e. a theoretical unit of power (equal to a megawatt) of a power plant that was never built thanks to energy savings (Lovins, 1997).

The American corporation PG&E (Pacific Gas and Electric) may serve as an example (Weizsäcker et al., 1999).

In the early 1980s, PG&E planned the construction of between ten and twenty nuclear power plants in California. The strategy was later changed: instead of building new power plants, PG&E focused mainly on energy efficiency projects targeted at individual customers, and renewables offered by small independent suppliers. The program was so successful that already in 1993 PG&E decided to close the department responsible for the construction of new power plants—it was no longer needed.

MHD (magnetohydrodynamic) generators are another example. MHD is a technology that converts heat directly into electricity. Gas is used as fuel; alternatively, it may be replaced by coal. The technology is based on the interaction between ionised gas and a stationary magnetic field (Kucowski et al., 1993). MHD power plants are about 15–25% more efficient than traditional coal-fired power plants and have a much lower environmental impact.

Nuclear power, which has gained more recognition in the recent years, is a more radical alternative. In Europe, nuclear power plants are located e.g. in France, Finland, Germany, Russia and Ukraine (Salay, 1997).

Two types of reactors are mainly used in the world today:

- PWR—Pressurized Water Reactors. They generate heat carried by water to a steam generator. Water is transported under high pressure to prevent water boiling in the reactor's cooling system. This technology accounts for 60% of the market.
- BWR—Boiling Water Reactors. Water plays a key role also in this type of reactors. Steam is generated in the reactor and used to drive a turbine. This technology accounts for 24% of the market.

The main problems in this area are as follows:

- Technical safety of the power plant and high costs of closing of the nuclear power units after the power plant is de-commissioned.

Table 39. Documented resources of coal in 2008 and the projected time of coal depletion, calculated as the amount of resources (R) to their annual extraction (E) vs. 2006 (World Crude Oil, 2011).

Region	Million short tonnes	Percentage of resources	R/E in years
North America	269,343	28.0	226
of which the USA	260,551	27.49	234
Central and South America	13,788	1.7	246
Europe and Eurasia of which:	335,566	35.0	237
Russia	173,074	18.26	no data
Poland	6,293	0.66	90
Middle East	1,326	0.3	194
Africa	34,934	4.0	no data
Asia and Oceania	293,042	31.0	85
of which China	126,215	13.21	48
The world	948,000	100	147

Table 40. Documented resources of natural gas in 2006 and the projected time of gas depletion, calculated as the amount of resources (R) to their annual extraction (E) vs. 2006 (World Crude Oil, 2011).

Region	Trillion cubic meters	Percentage of resources	R/E in years
North America	7.98	4.4	10.6
of which the USA	5.93	3.3	8.9
Central and South America	6.88	3.8	47.6
Europe and Eurasia of which:	64.13	35.3	59.8
Russia	47.65	26.3	77.8
Middle East	73.47	40.5	no data
Africa	14.18	7.8	78.6
Asia and Oceania	14.82	8.2	39.3
of which China	2.63	1.5	35.6
The world	181.46	100	63.3

Table 41. Documented resources of crude oil in 2008 and the projected time of oil depletion, calculated as the amount of resources (R) to their annual extraction (E) vs. 2006 (World Crude Oil, 2011).

Region	Billions of barrels	Percentage of resources	R/E in years
North America	211,559	16.0	12.0
of which the USA	21,317	1.6	11.9
Central and South America	109,857	8.3	41.2
of which Venezuela	87,035	6.43	77.6
Europe and Eurasia of which:	113,154	9.0	22.5
Russia	60,0	4.43	22.2
Middle East	748,286	55.5	no data
Africa	114,838	8.5	
Asia and Oceania	34,350	2.7	14.0
of which China	16,0	1.2	12.1
The world	1,354.19	100	40.5

- Management of the radioactive waste—the proposed waste management solutions (storing waste in closed underground mines, drilling of underground tunnels, or even ocean floor disposal) seem to be insufficiently safe.
- One cannot just switch off the plant after the accident. Only production of the electricity may be stopped, but cooling systems must still run for a very long time.
- Risks related to nuclear tests and the potential use of nuclear weapons during armed conflicts, especially in the Third World Countries.

The first controlled nuclear chain reaction took place in the world's first nuclear reactor in Chicago, the USA, in 1942 (Universal Encyclopedia, 2000).

Three years later, atomic bombs were dropped on Hiroshima and Nagasaki (6 August and 9 August 1945), marking a turning point in World War II, and at the same time causing an unprecedented environmental and social disaster.

Soon, the arms race between the West and the East began. As a result, nuclear power plants were built not only to generate electricity, but also to produce materials to be used in atomic bomb manufacture.

Technological safety proved to be the key element.

The explosion in unit 4 in the power plant in Chernobyl on 26 April 1986 (Jackson, 1996) and melting of the nuclear core was to a large extent related to the type of reactors used (RBMK, or channel-type graphite-moderated reactors).

The same event (core melt-down) was recorded on 29 March 1979 in Pennsylvania in the Three Mile Island nuclear power plant, but the installed safety systems prevented an explosion in this case (The Need for ..., 1979).

The main problem in RBMK reactors is the use of graphite as the neutron moderator. In other types of reactors, water is used as a moderator, and in the event of a water supply system breakdown, the chain reaction is stopped. In RBMKs the chain reaction is accelerated after the water supply is closed, leading to a disaster. As a result, this type of reactor was banned in many countries, including the United Kingdom (after a dangerous graphite stack fire in an RBMK reactor in an experimental stage).

In the Soviet Union, RBMKs were still built. The reason was twofold: both economic and military. First of all, RBMKs can use low-enriched uranium, which is much cheaper. And secondly, RBMKs generate large amounts of plutonium that can be used to produce nuclear warheads.

The exploding reactor in Chernobyl emitted large amounts of radioactive materials to the air—50,000,000 Ci according to official sources (equivalent of about 500 bombs dropped on Hiroshima), and 300,000,000 Ci according to unofficial sources (equivalent to 3000 bombs). In addition, the area within the distance of 30 km around the reactor was closed, and 30,000 people were evacuated and resettled (Jaskowski, 1988).

However, it was not the only such serious nuclear disaster in history.

In 1957, a tank holding 160 tonnes of highly radioactive waste from nuclear warheads exploded in the Kyshtym plant in Chelyabinsk in the Ural Mountains. The plant was located away from the European part of the Soviet Union, which made it possible to keep the disaster secret. It was officially revealed 40 years later. The area of 15,000 square kilometers was affected, of which at least 20% is still classified as highly radioactive (Borkowski, 1991).

The use of nuclear power in submarines is another problem.

Several submarines are known to have sunk as a result of a nuclear breakdown.

The most famous being the Russian submarine Kursk that sank on 12 August 2000. The wreck was cut into pieces and recovered during two operations in 2001 and 2002.

Another example is the case of the Komsomolets submarine that sank on 7 April 1989 in the North Sea while carrying nuclear missiles. In 1994, leaks of plutonium had been reported. Work to improve the tightness of its hull were undertaken in 1995–1996. The question remains: is it possible that radioactive substances will be ultimately released to the environment, causing nuclear pollution of the sea?

And what about nuclear weapons?

It must be noted that nuclear materials have slipped out of control. Nuclear tests also raise concerns. While watching the successful economic development of China, we should not forget the big picture—China's atomic bomb program. Information from Eastern Turkestan and the Sinkjang province are especially frightening. These are areas, with mountain peaks at 5000–7000 meters above the sea level. These areas were selected for China's nuclear tests—hidden behind the mountains that would stop the spread of radioactive substances to a large extent.

According to the available data, 19 nuclear tests were performed in 1963–1976, both in the air and on the ground. At least some of them were carried out in areas inhabited by people. All buildings, crops, and even livestock were destroyed. River beds were flattened out and water in rivers evaporated. Still, although this information has leaked, China faced no consequences on an international level (Lozinski, 1995).

Even the most state-of-the-art and peaceful technologies carry certain risks.

In 2002, a reactor in Denmark was closed due to safety concerns (Vince, 2002).

Earlier, in September 1999, a serious near-miss situation occurred in the Japanese power plant of Takaimura due to the non-compliance with safety procedures. The quantity of enriched uranium added to a container exceeded the technical standards—by as much as seven times. A chain reaction was started.

Luckily, it was stopped just in time. However, the increase in radiation levels was high enough to cause two deaths, and several hundred other staff members were exposed to excessive radiation (New Details on., 1999; Edwards, 2002).

In March 2011, after reaching 9 magnitude earthquakes, tsunami totally devastated Japanese Eastern Coastal Region. After the tsunami a lot of problems touched their nuclear power stations. The worst situation was in Fukushima: failures of cooling systems, dangerous explosions and radioactive leaks and nuclear fuel rods melting processes were reported! (Sky News, 2011). There were also problems with Onagawa and Tokai nuclear plants. 'Safe' reactors appeared to be not so much safe. It was the worst disaster in Japan since World War II and the worst accident in the history of western nuclear industry.

What is more, the first estimations are showing that recovery after this terrible havoc will cost Japan not less than $ 300 billion (Sky News, 2011).

These aspects should be considered when analyzing the perspectives for the future development of nuclear power (Moberg, 1986; Rosenkranz et al., 2006).

In this context, the use of renewable or 'alternative' energy sources is the most promising option. Renewable energy sources include:

- biomass,
- hydropower,
- solar energy,
- wind power,
- geothermal energy,
- biogas.[12]

In the case of biomass and biogas burning, as well as geothermal energy sources, emissions to the environment are generated (mainly NO_2 and SO_2 for biomass and biogas, and CO_2 and hydrogen sulphide for geothermal power), but they are much lower than in the case of fossil fuels (Boyle, 1996). For other renewable energy sources (RES), emissions are close to zero, although the equipment used to generate RES power is manufactured using metals and technologies that are not environmentally neutral.

Let us analyze the general conditions for the use of the environmentally friendly renewable energy sources:

- Biomass burning.
  Biomass materials include straw, peat, wood waste (from agriculture, forestry, or paper industry), as well as timber from special plantations. In Europe, Osier Willow (*salix viminalis*) is a popular source of biomass. Other sources include poplar (*populus*), Black Locust (*robinia pseudacacia*) and rose (*rosa multiflora*).
  The characteristic features of these species include (Mackow et al., 1993):
  - fast growth, even 10 times faster than the natural growth of biomass in an ordinary forest,

---

[12]This study focuses on the energy sector. Still, certain other solutions should be mentioned. There are also biofuels, a group that also includes liquid biofuels (such as bioethanol or plant oils). Liquid biofuels may be used in cars in addition to other new solutions such as hybrid engines or fuel cells. Development of these technologies is of prime importance, as the level of emissions of pollutants from transport vehicles using fossil fuels has been increasing steadily (Tengstrom & Thynell, 1997).

- high calorific value: two tonnes of dry timber equal one tonne of coal,
- no special requirements as regards soil and climate conditions—fallow land, land set aside, or even polluted or flooded land can be used,
- no special requirements as to temperature and climate—resistant to low temperatures,
- high resistance to pests and diseases,
- extended use of plantation (in the case of willows, crops are harvested every 2–3 years and the plantation may be used for up to 25 years).

- Hydropower plants.
  The main task of hydropower plants is to produce electricity. There is no need to produce steam as in coal-fired and nuclear power plants—hydropower itself drives the turbine that supplies energy to the power generator.
  From the technical point of view, there are four basic types of hydropower plants (Kucowski et al., 1993):
  - run-of-the-river plants (without a dam),
  - conventional plants (with a dam and reservoir),
  - pumped storage plants (reservoir is located above the plant; when the demand for energy is lower, for instance at night, the plant will pump water back to the reservoir to replenish the storage of water),
  - tidal power plants: using the power of tides to drive the turbine in both directions.

Also the OTEC plants (ocean thermal energy conversion plants) are an interesting solution. They are located in oceans near the equator, in the tropics, in Hawaii, Japan, or Indonesia. In this zone, the temperature of water is about 30°C at the surface, but only 7°C at the depth of 300–500 m. This difference is used as a source of heat converted to electricity.

The largest hydropower plant in the world is in China—the Three Gorges Dam on the Yangtze River. The plant was opened on 24 June 2003, but the full capacity of its storage reservoir was achieved in October 2010 (39.3 billion cubic meters, the reservoir stretches 700 km along the Mudong River). In 2008 it generated 80,8 TWh, and together with Gezhouba Dam the complex generated 97,9 TWh.

A second plant of this type is the Itaipu hydropower plant on the Parana river on the Brazil—Paraguay border, which in the same year generated 94,68 TWh (Wikipedia, 2011).

In many other countries there are no such powerful rivers as the Yangtze, but even small hydropower plants (below 10 or 5 MW) offer a number of benefits such as the following:

– electricity is produced without causing any pollution,
– protection by dams against flooding is provided.

There are also a number of environmental impacts that give rise to serious reservations:

– change of the local environmental conditions (especially hydrological), the landscape, or even the climate, which is an obstacle in the case of projects located near natural areas, like national parks, protected by law,
– obstacles to the natural migrations of fish,

– necessary relocation of inhabitants of human settlements that will be covered with water (in the case of the Three Gorges Dam on the Yangtze River, as many as 1.4 million people had to be relocated),
– problems keeping the optimum purity of water in storage reservoirs.

Therefore, the proper location of the dam is extremely important.

- Solar energy.
Solar energy is necessary for life on Earth. Living organisms are able to use only about 0.02% of this energy in a direct way. Indirect factors are also important. It is estimated that up to 50% of solar energy is converted to heat and used to heat up the planet, up to 30% is used as a source of light, and the remaining portion is a source of energy that guarantees optimal material flow in the form of such fundamental processes as water, carbon, sulphur, or coal cycles (Salay, 1997).
Solar energy may be also used to produce electricity or heat. There are two basic methods (Kucowski et al., 1993):
  o Heliothermic method—conversion of light energy into heat used to drive the turbine and power generating unit. Solar radiation is absorbed by the solar collectors. Collectors may be either flat-plate (heating up to about 90°C) or concentrating (parabolic, trough or dish—reaching 750°C in the case of dish-type collectors).
  o Helioelectric method—direct conversion of solar energy into electricity. From the technical point of view, this method is based on photocells (made of silica, gallium arsenide, or cadmium sulphide).
Solar collectors should be used in areas with sunny weather throughout the year in order to produce electricity from this source.
Photocells offer more flexibility—they can convert both direct radiation and dispersed energy (i.e. they can work even on cloudy days).
In most European countries, insolation is rather moderate and subject to high variations. Therefore, solar energy is used as a secondary source of energy in water heating systems or in heating systems in buildings. Space-based solar power plants (SBSP) would solve this problem; however, the transmission of electricity back to Earth is a technical barrier.
In another interesting project, the European Union is planning on building of a 400-billion-Euro solar power plant on the Sahara desert, which could cover about 20% of the EU's energy demand (Meinhold, 2009; Matlack 2010).

- Wind power.
The first windmills were built in Persia around 2000 B.C. In Europe, they were in use since late 10th-early 11th century (Nowak & Stachel, 2007). The second half of the 20th century made possible to use wind power to drive the turbines of electricity generating units.
Wind turbines are sometimes installed as standalone units to produce electricity for a single household. However, they are usually grouped in wind farms including tens or even hundreds of turbines. One of the biggest wind farms in the world is located near the town of Magdeburg in Germany, offering a capacity of 4.5 MW (Christen, 2005).
The greatest problem is posed by the high variability of wind conditions—both during the day/night and throughout the seasons. As a result, variations in the actual capacity of the wind farm cannot be avoided.

This problem may be eliminated by using more effective methods for the storage of electricity produced in periods when winds are stronger.

Another interesting solution is to generate hydrogen to carry energy. This 'carrier' could be used when winds are very weak or when there are no winds at all.

- Geothermal energy.
Geothermal projects use the natural heat of our planet—mostly in underground waters, but also energy stored in the Earth. Liquid magma is the main source of this heat. Natural decomposition of radioactive elements is an additional factor.

The point is to create an artificial cycle: boreholes are drilled to collect hot water, this water is used, and then pumped back to the Earth for re-heating.

The use of geothermal waters, initially for therapeutic purposes, has a long history. Geothermal waters were used by the Etruscans (who built special pools in the area of present-day Italy), the Maori (present-day New Zealand), and in Ancient Rome, India, and Japan.

Geothermal waters are gaining popularity as the source of energy used to produce both electricity and heat. The use of this RES depends to a large extent on the capacity of the source and temperature of water. Electricity can be produced at a temperature of 150–200°C.

The largest geothermal power producer is located in the USA—The Geysers (north of San Francisco) with 15 plants and a capacity of 725 MW (Geysers.com, 2010), while Iceland (Wikipedia, 2011) has the highest share of geothermal energy in a country's total energy balance—87% of heat and 24% of electricity (near all the rest of electricity, 75,4%, is from hydropower).

- Biogas.
Biogas is produced as a result of fermentation in municipal landfills, in wastewater treatment plants, or in purpose-built biogas plants (Pawlowska, 1999).

Biogas is composed mainly of methane and CO_2. Methane is particularly important, as it can produce both electricity and heat when burned.

Biogas burning generates a small quantity of pollution in the form of an additional emission of CO_2, but it reduces the emission of methane—a significant greenhouse gas with heat absorption coefficient much higher than CO_2.

As we can see, there are many ways to produce energy. However, will they be sufficient when the global resources of fossil fuels become scarce? This is one of the fundamental questions from the perspective of sustainable development, but it still remains unanswered.

Maybe a radical discovery of a new energy source will change the scenario. The energy of the Universe generated as a result of the Big Bang is one theoretical option (Michnowski, 2007). Is it possible that one day we will be able to capture and use it to the benefit of the entire human race?

Much will depend on our willingness and determination to search for and support the development of new technologies. In addition to financial support or political will, a solid legal system will be also necessary. Therefore, we will now move on to discuss another plane of sustainable development—the legal framework.

2 LEGAL PLANE

2.1 *Environmental protection and sustainable development law*

Law is a set of rules of conduct arising either from tradition or enforced directly through institutions. Sources of these rules and norms are different (Dider, 1992; Krapiec et al., 1992). They may be based on the 'natural law', the tradition of positivism, or legal realism.

The concept of 'natural law' dates back to ancient times. According to this concept, nature is governed by certain laws and it is our duty to identify and comply with them. The basic laws of nature include the right to live and self-development, and the overriding imperative is 'to do good'!

In contrast, the positivist approach focused exclusively on man-made laws.

In addition, the theory of legal realism states that the law is a specific set of social or psychological facts.

Norms change along with the ever-changing conditions of human existence. In ancient Mesopotamia, tribal customs were gradually abandoned as people started to move from their rural environments to cities. This led to the emergence of urban societies. According to Lewis Mumford, it was a source of many problems and ultimately led to the adoption of the famous 'Code of Hammurabi' that unified the Sumerian norms and laws, and helped Babylonia reach its peak in the 18th century B.C. (Redmann, 2006; Hall, 2001).

The key acts of law in the area of environmental protection and sustainable development are discussed in Chapter 1.

At this point, as part of our discussion of sustainable development levels, we will analyze the general conditions for legislative works. This is important, as these general considerations are easily overlooked in the plethora of documents and specific (both historic and modern day) solutions.

Let us note first that environmental protection and sustainable development law has a number of significant functions—see Table 42. Modern-day legal norms define environmental standards (e.g. the technical requirements, including the environmental impact assessment for new technologies and maximum emission levels).

Moreover, legal norms are formally mandated in:

- administrative law
- civil law, or
- criminal law.

Administrative law plays a significant role in this respect, as it governs the activities of public administration bodies which are key decision-makers in virtually all areas of social life. This body of law makes it possible to annul decisions that are harmful to the natural environment, to discontinue and prevent activities that have negative environmental impacts (usually business activities), and to impose an obligation to restore the natural environment to its former condition. At the administrative level it is also possible to impose penalties for non-compliances with the adopted laws.

Civil law, in contrast, makes it possible to get compensation for environmental damage. As such, it refers to the effects of human activity, not to their underlying causes—which may be seen as a weakness in some respects. It does happen that environmental damage is serious enough to be considered as environmental

Table 42. Functions of the environmental protection and sustainable development law (Paczuski, 2002; Runc, 1998).

Function	Description
Organization	Establish legal and systemic bases for the functioning of various entities (including the state).
Limitation and protection	Impose certain limitations on access to the environment (e.g. nature reserves or national parks).
Protection of rights of subjects	Protect rights in the form of compensatory claims (e.g. compensatory damages or repair of damage to property).
Business incentive	Introduce business incentive instruments, such as additional payments or subsidies.
Technical progress	Adopt technical standards and regulations.
Penalization	Establish sanctions for certain offences and misconduct. Penalties should have repressive and educational functions.
Protection of nature at the international level	Implementation of international laws by individual countries (conventions, agreements, declarations).

devastation, i.e. when nature is unable to restore its original condition (or this process will take many years and only on condition that the environmentally harmful activities are discontinued). If this is the case, the original condition of the environment cannot be easily restored regardless of the amount of money collected.

Criminal law is applied when any indictable offence or summary offence is committed.

In view of the foregoing, the environmental protection and sustainable development law should be focused particularly on the earliest possible intervention. It should be applied as soon as the activity that might affect the environment is commenced, or even at the planning stage when there are reasonable grounds to suspect that a given enterprise will have significant negative environmental impacts.

Enforceability of the adopted legal regulations is another problem. A number of different factors should be considered here (Matczak, 2000):

- Transparency of law. Legal regulations should clearly identify who they apply to, what activities are allowed, what restrictions can be applied, and for how long they will remain in force.
- Procedural safeguards referring to the enforceability of law.
- Consistency of legal regulations across various national legal acts.
- Consistency of national and international laws.
- Degree of knowledge of legal regulations among decision-makers.
- Relations with other social norms and standards (customs, moral norms).

In addition, legal norms should meet certain criteria (Boc & Samborska-Boc, 2005):

- Legal norms that set certain goals. These represent a general policy translated into the language of law and set the goals for public administration at the macro-level in specific problem areas.

- Technical directives in environmental protection standards. These refer, among other things, to spatial planning and environmental protection programs. The technical aspect is introduced into the legal framework by reference to a specific technical standard or even by including this standard directly in the directive.
- Discretionary environmental protection norms. These represent a flexible basis for the more effective implementation of environmental protection policies. They result from the fact that not all situations can be predicted and clearly defined within the existing fixed legal framework.
- Different geographical coverage. In general, legal norms are equally applicable to all subjects. However, it may happen that a given solution is strictly related to a given area and applies only to this specific area.

In addition, there are certain guidelines dedicated to the environmental protection laws (O'Riordan, 1981; Wrzosek, 1993):

- Equality of all entities as regards their rights and duties, also in the context of environmental protection (including equal fees paid for the use of the environment or equal penalties enforced).
- Preference for solutions that do not affect the environmental balance. This rule applies directly to the industry (e.g. the cleaner production concept described in the technical section of this book), and also to the standards of adoption of environmental decisions by public administration bodies.
- Co-operation between public administration bodies to counteract threats to the environment. This may take two forms. It can be part of the division of competencies between individual public administration bodies from the municipality up to the province level, or it can be in reference to the unification of methods adopted by different bodies at the same public administration level, but in different regions.
- Interdisciplinary approach to the formulation of acts of law. This principle is of key importance in the context of sustainable development.
- Adoption of environmental considerations in spatial planning, both at the level of local municipalities and in big cities (e.g. the 'green belt' or 'ecological corridor' concepts).
- Effective control of all types of activities, including business activities, in the context of their environmental impacts.

In practice, what is the current status of environmental protection and sustainable development law?

At this point, it would be worthwhile to review the key acts of law (described in more detail in Chapter 1) which form the overall legal framework in reference to global agreements and regional (international), national, and local laws. They all refer to the same underlying problems, but from a different perspective:

- Global agreements include solutions important for the proper functioning of the biosphere as a whole and refer to all countries throughout the world.
- The regional agreements focus on the consequences of trans-boundary pollution (in the case of bilateral agreements they will involve two countries only, but more often these agreements are signed by a group of countries).
- The national category includes solutions adopted by a given country and applying to its territory only.
- The last category is introduced in provinces within the country.

At the global level, agreements and conventions include documents adopted at the UN Earth Summits in Rio de Janeiro (1992) and Johannesburg (2002), as well as the "Kyoto Protocol" (1997). They come into force not when signed by representatives of individual countries, but only after they have been supported by the relevant national acts or regulations adopted by these countries (or, in the case of the European Union, by Community legislation).

At this point it should be noted that global conventions and agreements are usually the result of compromises, which often means that certain expectations as to the shape of the final document are not met.

For instance, according to scientists, the current agreements on the greenhouse gases emission limits (including "The Conventionon on Climate Change" adopted in Rio de Janeiro in 1992 or the "Kyoto Protocol" of 1997) are not restrictive enough to prevent possible negative climate changes, although they do introduce considerable emission limits (Ackerman, 2001). This problem will be discussed in more detail in the next chapter dealing with the political aspects of sustainable development.

At the international level, we should refer to the four main EU directives forming the framework of the integrated approach to environmental protection, concerning pollution prevention and control, air quality, water and waste—see Table 5.

On top of these, it should be noted the last two editions of the EU Environmental Action Programmes: the Fifth Environmental Action Programme, "Towards Sustainability", and the Sixth Environmental Action Programme, "Our Future, Our Choice".

Also, as part of the "Treaty of Nice", the European Union adopted the following environmental principles that constitute the synthesis of today's environmental protection and sustainable development legislation (Klemmensen et al., 2007):

- The Principle of Prevention—introducing the obligation to prevent pollution at source in production processes.
- The Principle of Early Consideration of Possible Environmental Impacts (the Precautionary Principle)—the decision-making process should start at the earliest stage of implementation of a given solution, e.g. a technical measure.
- The Polluter Pays Principle—costs of environmental protection and elimination of environmental threats should be paid by the polluter (see also Table 32).
- The Subsidiarity Principle—for all types of pollution, actions must be taken in proportion to the scale of the threats, that is at the local, regional, national, international, or the Community level (if necessary). In every case, a governmental task should be performed at the lowest possible level of government where it can still be done adequately (see also Table 45).

As a rule, these principles were placed in the context of actions directed at improving the quality of our lives, with the environment as the key factor for human development. This is in line with the idea of sustainable development.

In the context of national law, Poland may serve as an example. The concept of sustainable development has the highest status in the Polish law as it was included in the new Constitution, adopted in 1997.

Independently of the level (local, national, global), adoption of legal acts is always a source of heated debate, with a number of problems:

- Adoption of very detailed legal acts that lack a general overriding objective.
- A large number of new legal acts that is difficult to control. As a result, there is no consistency between individual acts of law.
- No consistency between the general acts of law and the resulting action programs, and insufficient consideration given to environmental protection matters in sector policies in certain countries (at the intersection of the legal and political dimension of sustainable development).
- Imprecise language of the legal acts that may lead to misinterpretations and legal loopholes.

To a greater or lesser extent, these problems affect all of today's legal systems. EU law is no exception to this rule. It is often criticized for being too extensive and inconsistent in some areas. It must be noted that by 2007 the EU has adopted 170,000 pages of directives, 59% of them within just one decade: 1997–2007 (The Daily Telegraph, 2007).

These directives have been subsequently extensively changed.

For example, the Waste Incineration Directive was amended four times in the span of just four years. We must bear in mind that changes in technology are expensive. If changes are too frequent, there is a constant risk of non-compliance with the ever-changing standards and it even may pose a threat to the industry (Christophe-Tchakaloff, 1999).

Another problem that concerns legal acts is that they contain too many insignificant details.

The requirements adopted for 'euro-cucumbers' are just one prominent example (Piontek, 2001):

- class 1 and class 2 cucumbers should have a small curvature (up to 1 cm per 10 cm of length, up to 2 cm per 10 cm for lower classes),
- minimum weight: 180 g for cucumbers cultivated in the ground, 250 g for cucumbers cultivated in greenhouses,
- in addition, cucumbers should be 'compacted' when packaged to avoid damage.

In practice, the actual quality of the product seems to be much more important than curvature, weight or size.

In the context of the inconsistency of law, it seems important that one of the key objectives of the Sixth Environmental Action Programme introduced in 2002 requires that the law must be made uniform.

Also, it should be noted that the question of consistency of the law has a historical dimension. The majority of today's laws originated in times when environmental considerations were virtually ignored and the concept of sustainable development did not exist at all.

2.2 Legal barriers to sustainable development

Law should refer not only to the corrections of existing documents or introduction of new legal acts. It has a more fundamental function: to identify the legal mechanisms that create barriers to the achievement of sustainable development.

These barriers are related to programs, or, in a wider sense, to doctrines that allow for or even encourage and promote activities that are environmentally unsustainable and are connected with the inefficient use of natural resources (Futrell, 2004). These activities give priority to business and social objectives rather than to other, especially environmental, considerations (Kassenberg, 2002).

Excessive subsidies received by obsolete and material-consuming industrial plants are just one example of this approach. Consideration should be given instead to upgrading or transforming them. The second option is usually more expensive in the short-term, but brings real benefits—economic and social, as well as environmental—in a longer term.

Also global corporations usually create barriers to the effective enforcement of environmental protection and sustainable development laws. As they usually operate on a global scale, the effectiveness of the legal systems adopted by individual countries, or even by communities, such as the European Union, is seriously undermined.

Food producers using the so-called 'factory farming' approach (i.e. huge livestock farms with thousands of animals each) is just one example of this practice. Their environmental impacts were discussed in the section dealing with the environmental dimensions of sustainable development. At this point, we will focus on the legal aspect.

Corporations usually claim that they operate within the boundaries of law (Frankel, 2004). In this context, let us now discuss a case in the state of Kentucky, where local authorities were virtually helpless when faced with a factory farm belonging to Tyson Foods Incorporated (Sierra Club, 2004). Although the company produced large amounts of waste, it denied any responsibility for environmental pollution caused by its activities—putting the blame instead on local farmers who acted as the company's suppliers. By this move the company made it much more difficult to clearly identify the source of the environmental pollution.

With the support of Sierra Club activists, a case was brought to court. In November 2003, the court decided that the corporation alone (not the local farmers) was responsible for environmental pollution. The settlement was reached in 2005. Under it, Tyson Foods Inc. was obliged to:

- Develop a system to precisely measure the level of emissions to the environment.
- Plant a belt of trees around the plant to create a natural barrier for odor factors.
- Pay damages to members of the local community who suffered as a result of the company's activities (damages to health or property).
- Pay damages to the Sierra Club to reimburse the costs related to the preparation and conduct of the legal proceedings (Sierra Club, 2004).

It may seem that the company's responsibility for environmental pollution caused by its activities was obvious, but corporations often find legal loopholes and their liability must be clearly demonstrated in a formal and legal manner. The Kentucky case served as a legal precedent that can be now applied to other factory farms. It became a practical tool to force corporations to comply with environmental protection.

All these factors should be considered when discussing the legal dimension of sustainable development, especially given that the law may be used to mandate certain solutions and to impose limits on other activities.

However, we must not forget that the law is adopted by politicians as decision-makers. Therefore, we will now move on to discuss the political dimension of sustainable development.

3   POLITICAL PLANE

3.1   *Policy and politics*

The terms 'policy' and 'politics' have a number of different meanings arising from different theoretical and methodological approaches (Lisicka, 2000).
These include:

- The activities of state institutions (formal and legal approach).
- The process of taking over and wielding power (behavioral approach).
- Decision making process as part of the political governance (rational approach).
- The function of a social system, referring to the resolution of conflicts and decision making as regards the division of goods and selection of interests (functional approach).
- A form of service to the society, the overriding objective of which is to solve problems related to the shortage of goods (post-behavioral approach).

In general, policy may be defined as "the art of ruling a country, region, municipality, or organization, i.e. a general activity of any authority focusing on the achievement or protection of certain objectives" (Poskrobko, 1998), including those connected with sustainable development. These objectives are defined in strategies and programs that constitute a platform for concrete actions.

Let us note that the terms 'strategy' and 'program' are used interchangeably. However, certain publications draw a distinct difference between the two:

- A strategy is a long-term vision with the time horizon of 20 to 30 years into the future.
- A program is a long-term policy with the time horizon of 10 to 15 years.
- An action program (sometimes referred to as an operational strategy) is a short-term policy with a time horizon of 1 to 4 years (Krikke & Zaworska-Matuga, 2001).

In practice, there are certain departures from this terminology, but usually the title of a given document defines the time horizon in years. What is important is that a program (or strategy) is usually based on underlying legal acts, but may also be used as a basis for the development and adoption of new acts of law.

The formulation of any sustainable development strategy is based on four stages (Andersson, 1997):

1. Identification of problems to be solved.
   This stage should be based not only on the proposals of scientists, but should also consider the external pressures, including those from NGOs, highlighting problems that are of particular importance for local communities

(Frissen, 2000). An inter-sector approach should be adopted here. Such an approach refers to all forms of business and social activity in a given area in the context of the long-term effects of any human activity. We must bear in mind that certain problems are shared by many different entities; therefore, it is possible to use solutions that have already been tested in other locations.

2. Formulation of the optimum strategy and consultations.

At this stage, it is important to adopt a realistic perspective; that is, to propose goals which are within the approved budget and are feasible given the existing conditions.

3. Implementation of the strategy.

4. Follow-up control.

This is a crucial stage, as it demonstrates that the strategy actually works in practice and not just on paper.

What is important is that the precautionary principle should be adopted from initial stage. This principle states that if an action has a suspected risk of causing harm to the public or to the environment, the action should not be taken (Andreasson-Gren, 1992; O'Riordan & Cameron, 1994; Mackenzie & Mackenzie, 1995; Verbruggen, 2000).

In addition, the decision will also depend on the current division of powers between political forces (Saint Marc, 1971).

A strategy may be (Kukulka, 1992) either active/preventive (i.e. trying to solve not only the existing problems, but also to prevent future threats) or reactive (merely adapting to the existing problems).

Other general principles are also important (Krikke & Zaworska-Matuga, 2001):

- Reference to previous policy.
- Focus on the future.
- Reference to the condition of the natural environment.
- Reference to the objectives to be achieved.
- Reference to the legal, financial, and social instruments that will be used to implement the strategy.
- Reference to the specific conditions in a given region.
- Social consultations involving people who will be affected by the strategy.
- The strategy must define the tasks of public administration bodies; at lower public administration levels, guidelines adopted under general programs should be considered (the local strategy should be in accordance with the overall national strategy).

One example of a global strategy is the proposal put forward by Al Gore, former Vice President of the USA under President Bill Clinton, in his book "Earth in the Balance" (2000). It is the 'Global Marshall Plan'. The name refers to the 'Marshall Plan' which the US introduced after World War II for the reconstruction of western European states. According to Gore, sustainable development will need a transformation comparable to the rebuilding of post-war Europe.

The main assumptions of this new plan can be summarized as follows (Andersson, 1997):

- Stabilizing the world population.
- Developing and making available new technologies.

- Introducing the new economic 'rules of the road', including the environmental costs of our economic activities and elimination of subsidies and grants from state budgets that are used to finance activities that have a negative impact on the environment.
- Developing new global agreement on the adoption and implementation of these goals.

This plan was devised in the days when Al Gore was still a member of the US Senate. As the US Vice President, he had to face the reality and the power of giant American corporations that are not too eager to make new technologies available to poorer countries, unless they pay the full price. In practice, the implementation of sustainable development strategies in the short-term is expensive, and so the proposed restrictions must be well grounded and justified.

Scientific research is the source of the necessary data. In this context, the relation between scientific papers and their interpretation that makes it possible to formulate recommendations under the adopted sustainable development strategy seems to be a major challenge.

Historical records indicate that often we would turn a blind eye to this data.

Sulfur smog is one of the oldest environmental threats known to humans (Sörlin, 1997). It became a problem for London in the 12th century when coal became the main source of energy for heating and cooking in households. In 1273, King Edward I banned the burning of coal in ovens in London (Wojciechowski, 1997). The first scientific analyses of the quality of London's air were performed in the late 18th century. The first concrete measures to control pollution were taken much later, in the 1950s (triggered mainly by the 1952 smog that killed about 4000 Londoners). As we can see, 700 years passed from the time when smog was first recognized as a problem until the moment when the first actions to combat this problem were taken! What is even more important is that these actions were finally taken as a result of media pressure, not as a reaction to scientific studies or to protect people's health (Sörlin, 1997).

Many other serious environmental threats were identified in the 19th century. The greenhouse effect is one example. In 1896, a Swedish scientist, Svante August Arrhenius, indicated that carbon dioxide emissions from industrial plants may be the cause of global warming; the results of his studies were not much different from today's research (Choo, 2006).

Also in the 19th century (in 1872), an English chemist, Robert A. Smith, presented the results of his studies on acidification of the environment caused by the emissions of sulfur oxides (Sörlin, 1997).

Despite their solid scientific backgrounds, the studies of Arrhenius and Smith were practically ignored. They became the subject of a heated political debate nearly 100 years later when the issue was brought up by the media. A Swedish chemist, Svante Oden, played a key role in the issue of the acidification of the environment. He presented the results of his studies to the Swedish magazine 'Dagens Nyheter' in 1967 (Sörlin, 1997, Choo, 2006).The media stirred up public opinion, and finally politicians had to face the problem.

Today, the relationships between politicians, scientists, and the media (especially TV and the Internet) play a major role. According to many people, the influence of the media on public opinion is a form of power. It is also a form of power in a political sense.

168

In this context it must be noted that the role of the media is appreciated not only by environmentalists.

The reaction of the tobacco industry in the mid 1950s to the disclosure of the relationship between smoking and the probability of developing cancer is a notable example. Tobacco corporations started an extensive disinformation campaign by sponsoring publications in the press to undermine the results of scientific studies (Robinson, 2007).

In recent years, similar actions have been taken in the context of the reduction of greenhouse gas (GHG) emissions to the atmosphere. Well, there are some scientific doubts connected with the greenhouse effect. However, in 2007 it was revealed that the US oil giant ExxonMobil (Gore, 2006; Robinson, 2007) spent as much as US $ 16 million in the span of seven years to finance 43 institutions and numerous publications, especially in the popular press. The goals of these disinformation attempts were to question the threats connected with the greenhouse effect, to divide public opinion, and to influence decision-makers. These attempts were successful—the disinformation campaign was one of the main reasons behind the rejection of the "Kyoto Protocol" by the USA. When these practices were finally revealed by the media, ExxonMobil radically changed its position when faced with the threat of consumer boycotts. It was a possible scenario at that time, because other oil corporations, including British Petroleum and Shell, had started to implement concrete measures to support more environmentally friendly technologies much earlier.

Brand image seems to be of utmost importance in this context. A corporation that supports an environmentally friendly policy enjoys a better reputation among consumers, which translates into better financial results.

It proved right for Wal-Mart, the largest department store network in the US. In 2006, the corporation announced the adoption of sustainable development principles for its business. The project started with the construction of two supermarkets with the following state-of-the-art, environmentally-friendly systems (Swiatek & Charytonowicz, 2006):

- Refrigeration (a large amount of energy was being wasted to cool the air around the fridge).
- Lighting (new energy-efficient solutions).
- Material and waste management (including a waste incineration plant that would produce thermal energy for in-house use).

The corporation also assumed the obligation to promote healthy foods produced by local farmers.

Wal-Mart's project is one good example, but it must be evaluated in a long-term perspective. There is a risk that in practice the project will turn out to be nothing more than propaganda; two stores, or even one network, are not enough to change the big picture—the entire chain of supermarkets must be involved.

The example of Exxon Mobil raises the following question. How many other corporations have tried to take such negative actions and how many cases have not been revealed yet? These are the key factors that will determine whether a given strategy is just a nice colorful document—a 'good on paper' solution—or a truly effective measure.

In western democracies, numerous non-governmental organizations (NGOs) ensure objective and independent assessment. These include World Watch

Institute, Human Rights Watch, Sierra Club, or the most famous NGO—the Club of Rome—to name but a few.

However, before any strategy is put into practice, it needs political will. The situation differs by country—see Table 43.

In the case of the more developed EU member states classified as 'cautiously supportive', we should note that the evaluation was performed after the conclusion of the Fifth Environmental Action Programme, "Towards Sustainability", but before the launch of the Sixth Environmental Action Programme, "Our Future, Our Choice" (Ryden et al., 2003). The sixth edition unified a number of policies, which is an important factor that could have raised the score.

At the same time, the political future of the European Union is by no means obvious. There are a number of alternative scenarios, including both the strengthening and weakening of the Community—see Table 44. Will the 'Union modernizing its peripheries' scenario become a reality? Hopefully it will, but it remains to be seen.

However, the effects of the political will sometimes may be contrary to all good intentions.

Table 43.   Support for sustainable development in selected industrialized countries (Lafferty & Meadowcroft, 2000).

Enthusiasts	Cautiously supportive	Disinterested
Netherlands	Australia	USA
Norway	Canada	
Sweden	Germany	
	Japan	
	UK	

Table 44.   Scenarios of the European Union's political future (Lafferty & Meadowcroft, 2000).

Scenario	Description
Union as a fortress	Defending itself against globalization, entrenched within the Schengen zone.
Union as a colony	Dominated by the Anglo-Saxon global capitalism model that might ultimately consume the Community.
Union of homelands	The Union created by separate nations in its current shape.
Union of tribes	Tribal Europe characterized by increasing tension and conflicts between nations.
Union as network Europe	Europe as a diversified network established on the basis of today's countries transformed into new territorial and social entities.
Union of regions	Europe consisting of sub-national regions.
Union as an empire with a centre and peripheries	Intensive development of the 'old' civilization centers of the Community, gradually accompanied by the western centers of new members from the East.
Union modernizing its Peripheries	Elimination of civilization gaps between individual areas of the Community—truly sustainable development.
Union as a bureaucratic utopia	Excessive bureaucracy halting the Community's future development.

For example, in order to eliminate the problem of environmental pollution caused by the rapidly increasing number of cars, an order was adopted in Mexico that one registered vehicle only may be used on one selected day of the week. The expected drop in road traffic was never achieved—the inhabitants simply bought more cars and owned several vehicles—a different car would be used every day (Connolly, 1999). In this case, the political decision was based on environmental considerations, but the equally important social dimension was simply disregarded.

A successful strategy of sustainable development must, therefore, consider all dimensions, and in particular it should propose a feasible implementation of all objectives with due consideration to the traditions and customs of specific local communities.

3.2 Democracy and sustainable development

The discussion on sustainable development should also be placed in the context of the general principles and rules adopted by democratic societies, where the common good is an overriding objective of the system—as the concept was defined in ancient Greece (Jaroszynski, 1995)—and where all stakeholders (public administration, NGOs, and individuals) are represented. Democracy gets criticized at times, but, as Sir Winston Churchill once said, "Democracy is the worst form of government except all those other forms that have been tried from time to time" (Weizsäcker et al., 1999). The selected principles of democracy are presented in Table 45.

These central tenets of democracy are followed by democratic countries[13]. The question is: do democratic countries also follow these principles in their relations with non-democratic ones?

Table 45. Principles of democracy (Andersson, 1997).

Principle	Description
Equality	Acceptance of both benefits and costs by the interested parties. Inequality occurs when one party bears all the costs without gaining any benefits.
Subsidiarity	Action must be taken at the lowest level of governance possible. Local authorities have the best insight into the specific problems of their region.
Responsibility of decision makers	Decisions taken by authorities must be based on clearly defined criteria, must be made publicly available, and cannot be influenced by private interests.
Social participation	Social consultations should precede any decision making process. This is in line with the concept of a 'civic society' promoted by democratic countries.
Common development	By influencing the relationship between the authorities and society, citizens develop not only themselves, but also the mechanism of democracy.

[13] However, some problems are identified in democratic countries, too. The USA is the one which has the highest percentage of its population in prison, with significant disparity between the numbers of white and Afro-American prisoners as much as 1:6 (Human Rights Watch, 2007).

This question is justified particularly in the context of the development of today's global market. Democratic societies usually support the free market economy, but a free market does not necessarily support democracy as such (Barber, 1995). We should also note that the free market economy was adopted by many countries where democracy is restricted to a large extent. The fast-developing economy of China is a prominent example—the country is sometimes referred to as the least democratic country in the world (Barber, 1995). Moreover, the report of the NGO Human Rights Watch published in January 2008 clearly shows that 'old' democracies (such as the USA or the European Union) often turn a blind eye and deaf ear to the acts of terror, violation of human rights (both civil and political), lack of democracy, and election fraud committed by their trade partners. In other words, they accept it. China is not the only non-democratic country in the spotlight—it is accompanied by the likes of Pakistan, Chad, Columbia, Congo, Ethiopia, Kenya, Iraq, Somalia, Sri Lanka, Sudan, Cuba, Libya, Iran, North Korea, Saudi Arabia, and Vietnam.

Concrete examples can be given.

For instance, European countries bought illegal timber from Liberian rebels, who then used the money to buy weapons. Buying products manufactured by children, usually in slave-like conditions, is also widely condemned. It turns out that not so long ago, the United Kingdom was using coal in its power plants that was extracted from Columbian mines by children (Hermitte, 1999)!

Irrespective of the actual level of democracy, there is not one country in the world that is completely independent of other countries (Dasmann et al., 1980). That is why international agreements, often under the auspices of the United Nations, are so important.

Agreements relating to the protection of the ozone layer by reducing emissions of freons and halons that contain ozone-depleting chlorine and bromine are an important example (Thornblad, 1992). These agreements were adopted under the United Nations Environment Programme (UNEP).

The first document was signed in Montreal in 1987—the "Montreal Protocol on Substances That Deplete the Ozone Layer" (Meadows et al., 1992). The protocol assumed that the production of freons would be halted at the level recorded in 1986, and subsequently reduced by 30% in 1993. Promotion of technologies that do not contain any freons was also advocated.

The protocol was not a complete success, as most of the third world countries refused to sign it for fear of the high costs connected with the introduction of new technologies.

China is another example. It rejected the protocol because the country had just started the mass production of refrigerators that used freons—mostly for its domestic market. Given the scale of its production and its population, decisions taken by China have a global, not just a local impact.

A year later, it was discovered that the ozone was being depleted much faster than expected. As a result, the next round of negotiations was launched in 1990 and closed with the signing of the "London Agreement" (Meadows et al., 1992). A more restrictive plan for the reduction of freons was adopted. What was equally important, a special fund for the third world countries was established to help them implement technologies that do not contain chlorine or bromine.

The next several years have shown that the London Ozone Agreement proved effective—emissions of freons to the atmosphere dropped significantly.

However, the ozone hole threat has not been eliminated so far—it may take up to 100 years to remove chlorine from the atmosphere (Meadows et al., 1992).

Not all agreements are successful. "The Kyoto Protocol" to "The United Nations Framework Convention on Climate Change" (adopted in Rio de Janeiro in 1992) is the most famous example. This protocol was negotiated in December 1997 and assumed a reduction in greenhouse gases of 5.2% on average by 2012 (vs. the 1990 baseline). For the protocol to come into effect, it had to be signed by at least 55 countries responsible for at least 55% of the global carbon dioxide emissions (Kyoto Protocol, 1998). This goal was not achieved for a long time—mostly due to resistance from the USA and Russia. Finally, Russia ratified the protocol in December 2004, and it became effective in 2005 (Kyoto Protocol, 2007).

That the United States has not endorsed the protocol is particularly important. The USA is placed second—with a 19.91% share—in a ranking of countries by emissions of CO_2 to the atmosphere (CDIAC, 2010).

As a result of the nearly 10-year delay in the adoption of "The Kyoto Protocol" the actual emission of anthropogenic greenhouse gases was reduced by a much lower amount than assumed. The period originally adopted (reduced emissions by 2012) had shrunk.

Therefore, another conference was held on December 3–15, 2007 in Bali, Indonesia. This conference was attended by the representatives of 180 countries (UNFCCC, 2007). The goal was to extend the time horizon of the Kyoto Protocol and to considerably reduce the level of GHGs emissions from 25% to 40% (vs. 1990) by 2020. Both these attempts failed. An agreement on the necessary action against climate change was finally signed, but included no concrete proposals. Again, objections from the United States tipped the scales (Bryner, 2000). The next conference in Stockholm in 2009 provided no breakthrough either.

However, the lack of political will to reduce greenhouse gases emissions from the US administration was no obstacle to a number of environmentally friendly initiatives of the American Environmental Protection Agency (EPA) or to a number of regional agreements in the USA, the most notable of which included (Grossarth & Hecht, 2007):

- Regional Greenhouse Gas Initiative (RGGI), signed by 10 states representing more than 46 million US citizens. The initiative covered the states of Maine, New Hampshire, Vermont, Connecticut, New York, New Jersey, Delaware, Massachusetts, Maryland, Rhode Island. Few more states have the status of observers (RGGI, 2011).
- The undertaking to reduce GHG emissions signed by individual cities. The list includes more than 50 agglomerations, such as Atlanta, Baltimore, Berkeley, Boston, Chicago, Ohio, Denver, Las Vegas, Los Angeles, Miami, Minneapolis, Nashville, New Orleans, New York, Philadelphia, Pittsburgh, Salt Lake City, San Francisco, Seattle, and Vancouver (Choo, 2006; US Conference, 2007).

Efforts by the US politician Al Gore also deserve recognition. In recent years, he has become famous for his involvement in actions targeting reduction of greenhouse gases emissions mostly from developed countries. In recognition of these efforts he was awarded the Nobel Peace Prize in 1997 together with the

Intergovernmental Panel on Climate Change (IPPC). It may seem paradoxical that this prestigious prize was awarded to a politician for his actions that were against the then official policy of his own country.

In addition to his meetings with politicians and scientists, Al Gore organized a number of spectacular projects in the media. He starred in a documentary movie "An Inconvenient Truth" which won an Oscar for the best documentary in 2007. The movie was accompanied by a book under the same title (Gore, 2006) and a series of "Live Earth" benefit concerts, the goal of which was to direct the attention of societies all over the world to the problem of global warming. Performances were given by Genesis, Duran Duran, Red Hot Chili Peppers, Metallica, Madonna, Geri Halliwell, Roger Waters, The Police, Bon Jovi, UB40, Katie Melua, Cat Stevens, and Lenny Kravitz. The event was held July 7, 2007 simultaneously in Sidney, Johannesburg, New Jersey, Rio de Janeiro, Tokyo, Kyoto, Shanghai, London, Hamburg, Washington, Rome, and Antarctica! These venues were selected on purpose—the list included cities hosting the Earth Summits on sustainable development (Rio de Janeiro, Johannesburg), directly connected with the greenhouse gases emission protocols (Kyoto), or most threatened by the climate changes (Antarctica).

Al Gore appeared on stage in person, asking the audience to support the 'Seven Point Pledge':

- To demand that their countries join an international treaty that cuts global warming pollution by 90% in developed countries and by 50% worldwide.
- To take personal action to help solve the climate crisis by reducing its own CO_2 pollution, so to be 'carbon neutral'.
- To fight for a moratorium on the construction of any new generating facility that burns coal without the capacity to safely trap and store CO_2.
- To work for a dramatic increase in the energy efficiency of their homes, workplaces, schools, places of worship, and means of transportation.
- To fight for laws and policies which expand the use of renewable energy sources.
- To plant new trees and join with others in preserving and protecting forests.
- To buy from businesses and support leaders who share a commitment to protect the environment.

The 'Seven Point Pledge' is in accordance not only with the policy to combat climate change, but also with a number of strategies adopted to fight excessive consumption.

This aspect was also highlighted by the European Union. The concept of sustainable consumption was defined by the Council of Europe in 2003 as one of the key priorities in reference to the implementation of the postulates of the Earth Summit in Johannesburg. In addition, many EU Member States adopted concrete strategies in this area (Wasilewski, 2006):

- Germany, Finland: the "National Sustainable Consumption and Production Process".
- The UK: the "Sustainable Consumption and Production Action Plan and Market Transformation Programme".
- Austria: "Eco-Efficiency Programme".

Will these initiatives prove successful?

To a large extent, it will depend on the habits of consumers and these are not easy to change.

Problems with the feasibility of these postulates are used as an argument against documents adopted at the Earth Summit in Rio in 1992. Opinions have been voiced that the United Nations is unable to implement a sustainable development program in any real way.

Indeed, the UN has lost its leading role in recent years. The war declared on Iraq by the USA with virtually no regard to UN procedures is just one example. There are even voices saying that the UN should be replaced by a new organization, or even a 'world government' equipped with special global instruments to wield its power (Tinbergen, 1978).

In my opinion, these ideas are not realistic. Also, instead of removing the existing structure, and building anew, it is better to upgrade it. One proposal is to create a new UN agency, the World Sustainable Development Organization (WSDO), with the authority to impose sanctions on countries that do not follow the recommendations (Annan, 1997; Gupta, 2002).

It remains a matter of opinion whether this organization (or any other, for that matter) acting within the UN (or not) would be able to change the consumption patterns in developed countries (Pfahl, 2005). Maybe the implementation of new environmental education programs would prove more successful.

But not only negative aspects of policy should be brought into the spotlight. For example, despite all its shortcomings, "Agenda 21", formulated by the UN, is a unique instrument that helps build sustainable development strategies at many different levels—global, regional and local. The feasibility of any strategy depends on its authors—their knowledge and the political pressure they can exert. It is particularly important from the regional and local perspective, where the consequences of any action can be seen almost immediately (Buckingham & Theobald, 2003).

The local sustainable development program is a practical application of a famous saying of the environmentalists: 'Think globally, act locally'. It is also the practical implementation of the principle of subsidiarity that states that any political decision should be taken at the lowest administrative level possible, and most preferably locally (Andersson, 1997; Williams & Dair, 2007).

Therefore, we must always bear in mind the specificity of a given administrative unit (municipality, province, or region) that will be discussed in any document. Recommendations implemented in one region may prove totally unsuccessful in another.

The issue of local variability, which makes it difficult to apply universal one-size-fits-all solutions, is not the only problem encountered when developing general political strategies for sustainable development. Other problems that must also be considered include (Andersson, 1997; Kozlowski, 2005):

- The necessary interdisciplinary character of the strategy applying to all dimensions and levels of sustainable development. The group of specialists that will prepare the strategy should also be interdisciplinary—it should even include philosophers (see chapter 3)! Otherwise, we may end up with another standard environmental protection or nature conservation program instead of a comprehensive sustainable development strategy.

- The tendency to adopt strategies that are not feasible in practice. A 'colorful' strategy always looks good on paper. For instance, the best way to improve the quality of water might be to introduce an obligation to locate water intake points used by an industrial plant downstream from its effluent discharge point (Poskrobko & Olenska, 2001).
- It is fairly easy to be in opposition to environmentally friendly investments, as they usually require substantial financing as soon as possible, but their effects will be visible only after a longer time (Dasmann et al., 1973)—see Table 46. As a result, ad hoc strategies are developed that will not lead to any structural changes and will bring no benefits in the long term.
- The focus on long-term objectives does not mean that current problems are ignored in particular in areas other than environmental protection. Otherwise, we will go from one extreme to the other (Pawłowski, 2007).
- The time horizon that politicians usually focus on refers to current affairs in the country and the time left to the next election, i.e. a maximum of 4 years. As they want to be re-elected, political leaders usually steer clear of any unpopular (e.g. costly) solutions, even if they are necessary in the long term.
- The fact that the strategy must go through all levels of the decision-making process imposes additional limitations. A block at just one level is sufficient to put the entire strategy on hold.
- In addition, every strategy is under pressure from various interest groups. They focus on their own business, not the common good of society.
- On top of that, it is sometimes difficult to persuade ordinary citizens to accept and follow a strategy, even the best one, especially if the strategy is imposed by external institutions. Local consultations are, therefore, necessary. All too often, strategies are developed without even trying to include the inhabitants of the affected region in the process.
- It is important to seek support among representatives of local elites. Good word of mouth support often means much more than a long list of arguments.

The discussion should consider another important election instrument referred to as 'vote with the wallet' (Andersson, 1997). As consumers, we must

Table 46. Short- and long-term impacts of environmental problems and their scale: geographical coverage (Matczak, 2000).

Coverage	Hours and days	Months	Years	Decades	Centuries
	Time scale of the effects of environmental problems				
Local	Smog		Excessive eutrophication of waters	Waste deposition	Re-cultivation of soils
National		Floods		Wildlife extinction	Deforestation
International (continental)			Droughts	Acid rains Desertification	
Global			Ocean pollution	Ozone depletion Greenhouse effect	Lack of biodiversity

all decide which products we will buy and which we cannot afford. The same applies to the choice between environmentally friendly and environmentally harmful products.

However, regardless of the decision making level, the consequences of any decision will go way beyond the political dimension. All levels of human activity are interrelated, and sustainable development cannot be achieved in isolation in one area. But it may be possible if all levels are fully integrated. This problem will be discussed in the next chapter.

Sustainable Development as a Civilizational Revolution – Pawłowski
© 2011 Taylor & Francis Group, London, ISBN 978-0-415-57860-8

CHAPTER 6

Integration of planes, the phenomenon of globalization and the Sustainable Development Revolution

1 OVERLAPPING OF SUSTAINABLE DEVELOPMENT PLANES

When in the late 19th century electric lights replaced gas lamps, the air quality in cities instantly improved. However, in the second half of the 20th century the development of coal-fueled industries led to much greater air pollution, which spread far beyond the urban environments and became a global issue. It was a peculiar paradox. The solution that brought short-term positive effects has grown to be a threat in the long run (Fox-Penner, 1997). Also today many innovations which are meant to bring improvement might actually pose a threat.

For example, the increasing automation of production processes reduces their cost, but at the same time it generates unemployment. Fewer and fewer people are needed for the production of more and more goods. For instance: one machine at a coalmine, operated by just 5 people, extracts thousands tonnes of coal a day (Seidler & Peschke, 2004).

This is problematic even in such countries as Sweden where there is a general awareness of the need for environment protection, and the concept of sustainable development is widely discussed. In the late 1990s, hybrid-powered buses (internal combustion engine and electric motor) were introduced in Stockholm. They pollute the environment to a much smaller degree than the standard passenger vehicles. The next step intended was the construction of automatic buses travelling along specially devised routes. In this case, drivers would have become redundant. Can such a solution be regarded as sustainable? Viewed at most levels, the idea seems appropriate. It is hard to find fault from an ecological point of view (lower emission of pollutants and thus a 'healthier' environment), from a technological level (application of advanced, environmentally friendly technologies), from an economic level (the solution is cheaper than the standard one in the long-term perspective), and from a legal viewpoint (no regulations have been violated). However, there is also the social perspective (possible unemployment) and the related level of moral assessment of the entire project.

It is even more difficult to evaluate projects implemented on a greater scale. Help might be provided by the concept of the 'ecological footprint', developed within the framework of the idea of sustainable development. It is the area necessary to provide the resources and materials essential to the life of a given unit—a city, society, or any population under study (Wackernagel & Rees, 1996; Moffatt, 2000). The estimate of the ecological footprint is based partly on analytical and partly on hypothetical data. It addresses such issues as the area of forest which

179

should be planted to absorb the CO_2 emitted by the unit under study.[14] The calculations indicate that even in the case of a medium-sized city, the ecological footprint might be 500 to 1000 times the area occupied by the city itself.

And what is the situation in global terms? Taking into account the available land resources, there should be about 1.8 hectare of ecological footprint per person.

There are regions (of the symbolic South) where people need much less, but there are also areas (in the North) where the situation is the opposite and a much greater amount of land is required per person.

The world average footprint is now 2.7 ha—see Table 47. This means, that we need 1.5 planets to cover our needs, but we have just the one planet in reality (Living Planet Report, 2010).

The data confirm the thesis stated here in the discussion of the social level of sustainable development, that it is not the excessive population growth in the South, but the excessive use of natural resources by the countries of the North, which is the principal cause of contemporary environmental threats (Lipton, 1997).

With reference to this, we ought to address a specific subject, namely, climate change. In accordance with the precautionary principle, the specific solution to this problem would be to limit the emission of greenhouse gases to the atmosphere, but the introduction of such an arrangement has been resisted by many countries.

Table 47. Ecological footprint for different regions and selected countries on Earth in 2007 (Living Planet Report, 2010).

Country/region	Ecological footprint (ha/person)	Country/region	Ecological footprint (ha/person)
Africa:	1.41	Latin America and	
Mauritius	4.26	the Caribbean:	2.58
Central African Republic	1.32	Mexico	3.00
Malawi	0.73	Puerto Rico	0.04
Asia:	1.78	North America:	7.90
United Arab Emirates	10.68	Canada	7.01
China	2.21*	USA	8.00
Oceania:	5.39	Europe:	4.68
Australia	6.84	Denmark	8.26
New Zealand	4.89	Germany	5.08
		Ukraine	2.90

World average: 2.70 hectare, critical point: 1.8 hectare

* However this is distributed over the biggest population in the world (1,336,550,000).

---

[14]This forest does not actually exist. It is paradoxically a weakness, as it introduces a virtual factor, but also a strength of the concept, as it enables a more convincing illustration of a unit's environmental impact. A more important is the lack of differentiation between areas in which the environment is degraded, and those which are only reasonably exploited (den Bergh & Verbuggen, 1999).

The issue is also closely related to the mass destruction of tropical forests. Their logging contributes to global warming (perhaps to an even greater extent than greenhouse gas emissions) and leads to climatic anomalies which may occur anywhere on our planet. Moreover, it is extremely difficult to restore the ecological function of areas where the tropical forest had been destroyed. The circulation of matter is very fast in these areas, and as a result of subsequent agricultural activity the soil is quickly degraded.

Tropical forests are the wealth of the South (Myers, 1986; Lean & Hinrichsen, 1992). They would not have been cut down on such a massive scale if it was not for the rich countries' demands for wood. This way, the natural and ecological level (wealth of tropical forests) meets the economic and social ones (destruction of the forests in order to earn a living). If the inhabitants of these countries saw any other alternative, the forests could have been saved.

In the meantime, the climate changes are progressing. A prelude to them took place in the summer of 2002 when the residents of the small island of Shishmaref, situated off the coast of Alaska, decided to leave their homes. The direct cause of this was the rise of the sea level due to polar glacier meltdown and permafrost thawing. The island is in danger of total inundation. However, it turned out that the residents could not afford to finance the move of the whole island population. The estimated cost was US $ 150 million. Therefore, despite the threat, they decided to stay in their homes (Choo, 2006) and fortunately, obtained some funds to construct seawalls which can protect at least part of the shoreline (Wikipedia, 2011).

After all, Alaska is not among the poorest regions of the world. What should the poorer people living in densely populated areas do, when, their homes are also threatened with flooding?

Tropical deforestation and greenhouse gas emissions are actions taken locally, always at a specific place, but their consequences are global. A number of other forms of human activity are also characterized by such a broad spectrum of impacts. It is thus impossible to discuss the interpenetration of all levels of sustainable development without taking into account the global dimension, and specifically the process of total globalization which has never occurred before on such a scale in the history of mankind.

2   THE CHALLENGES OF GLOBALIZATION

The first historical form of globalization was the emergence of empires conquering other countries and introducing their customs, cultures, and religions into them. Such processes occurred from around 8000 BC to around the year 1500 (Wallerstein, 1999). Illustrations of such events are the empire of Alexander the Great and the Roman Empire.

The next phase of development was the colonial empires. Their emergence resulted not only from the enormous military power of some countries, but also by important geographical discoveries, particularly Vasco da Gama's discovery of the route to India around Africa (1498), and Columbus' arrival in America (1492). The first great empires were developed by Portugal and Spain. Later came those of France, Netherlands, and England. Throughout the centuries the ownership of colonies has been changing, but, nonetheless, colonialism survived until as late as the mid-20th century.

World War II was an important point in time, which, while in most cases it did not affect the colonies themselves, it did weaken the colonial powers.

Another breakthrough came with the resolution of the UN General Assembly which disapproved of colonialism and granted all peoples the right to independence. The last empire, owned by Portugal, fell as late as the 1970s (although if one regards the USSR as an empire, it ceased to exist even later—in the 1990s).

However contemporary globalization does not depend on the power of individual countries.

The term 'globalization' appeared in economic literature in the 1960s and referred to the activity of large international corporations (Micklethwait & Wooldridge, 2000; Bowman et al., 2006; Gawor, 2008). The idea itself arose much earlier. In the 1940s, American corporations often invested abroad thus bypassing trade barriers and capitalizing on the lower costs of production arising from, among other things, lower wages (Mucha-Leszko, 2005).

Today globalization is defined (Gawor, 2008) as an integrated, world-wide social and economic system related to large corporations, characterized by supra-state (and supra-national) diffusion of capital and the adoption of the free trade principle in the field of economic globalization and the assimilation of cultural models, especially in the mass form (cultural globalization).

Globalization may be based on egoistic axiology or eco-humanistic axiology—see the comparison of the two in Table 48. The latter, discussed as 'inclusive globalization' (Annan, 1997; Michnowski, 2004) though rooted in the principle of common good and compliant with sustainable development, is not the one which prevails. It is because the world of modern corporations is based on egoistic globalization, the priority of which is the uncompromising pursuit of maximum profit, where morality and ethics do not exist (Weizsäcker et al., 1999).

Globalization applies to all levels of sustainable development.

The economic level is the starting point. This is because globalization is based on major changes related to world trade, or, more broadly, to the flow of capital. This process was defined by Edward Luttwak as 'Turbo-capitalism' (Luttwak, 1999). It illustrates the rapidity and scale of occurring changes. The corporations want to earn more and more, so they promote consumerism in order to sell their products. Let us consider one of the characteristics of the process. Although there are plenty of estimates referring to, for example the profits of pharmaceutical companies in relation to the pharmaceuticals sold, it is difficult to find the results of objective research which studies the real effect of the pharmaceuticals sold on patients' health (Michnowski, 2007).

Consumerism is particularly targeted at young people, or even children. It is enough to say that advertisements targeted at children in the American market are worth US $ 2 billion a year (Goodwin, 2001). These actions are effective. American 3-year-olds are not yet able to write, but they can correctly pronounce the names of corporations or specific products and ask their parents to buy them (Goodwin, 2001).

A significant trait of global consumerism is also the unification of consumer expectations, e.g. the same fashionable clothes, the same electronic equipment, the same popular culture, the same beverages (cultural globalization). For instance, one of the most famous global concerns, McDonald's, daily serves over 20 million clients—more than the entire population of Greece, Ireland, or Switzerland. It is more than a coincidence that, for example in Japan, the most

Table 48. Globalization and sustainable development (Piontek, 2002, 2003).

Issue	Eco-humanistic inclusive globalization	Egoistic globalization
Position of man	Priority of man over other capital.	Human value depends on their economic usability.
Quality of life	The resultant of economic, social, menvironmental, and spiritual needs.	Only economic wealth.
Time perspective	Present and future generations.	Only the present generation, or more precisely, that part related to the 'global elite'.
Entity responsible	National state.	Anonymous supranational powers.
Law constitution	Aim: international law should be coherent with sustainable development.	Relativised legal system.
Role of the market	Fair market.	'Free' market based solely on economic efficiency.
Social divisions	Aim is social balance: equal access to knowledge, innovation, and technology. Emphasis on human dignity.	Clear division into the poor, the rich, and the super rich global elite.
Interpersonal relations	Emphasis on the significance of proper family and social relations for the correct mental development of man.	Subordination to the criterion of economic efficiency, weakened family and social relations.
Model of consumption	Limited consumption model.	Unlimited consumption model.
The problem of unemployment	Maintaining proper proportions between labor-consumption, capital-consumption, and environment-consumption.	Lack of will to solve the problem, it does not concern the global elite.
Environment protection	Inseparable element of the development process.	Variable dependent on the given combination of the 'free' market, competition, and profit rate.

popular network of restaurants in terms of the number of clients is none of the local chains, but McDonald's (Barber, 1995).

Consumerism is driven by both production growth and the development of newer and newer versions of products and devices. Thus, corporations support technological progress (the technological plane).

However, the growing production quickens the rate of consumption of the resources necessary for the production processes and is responsible for the general intensification of environmental pressure (the environmental plane). It is estimated that the literature already contains descriptions of over 10 million substances synthesized by man, and each year about a thousand new ones are introduced on the market. Unfortunately, according to some sources, only a small number of them (e.g. in the USA just 2%) have been comprehensively analyzed with regard to their influence on human health (Backlund et al., 1992; Bruyn, 2000). This is because the existing laboratories are not capable of completing more research and the market pressure for new products is immense.

Furthermore, even the positive aspects of technological progress may be wiped out by consumerism. This is true for example in the case of the automotive industry. Cars are getting better and better—more environmentally friendly—but they are also growing in number, so the total amount of pollution emitted is not decreasing and in many areas it is actually growing.

Another important issue is the acquisition of smaller companies by larger corporations, which results in significant restructuring.

One example is the Pacific Lumber Company which is involved in the cutting of sequoia trees. It adopted important environmental and social assumptions (Korten, 2007):

- The level of logging must not exceed the forest's regenerative capacity.
- The employees not only received remuneration, but were covered by a social program including a pension fund.

When the company was taken over by a larger entity, the logging drastically increased and the social benefits of employees were just as drastically limited (Korten, 2007).

This example shows that the global free market is not capable of solving environmental or social problems.

Moreover, corporations not only shape the operations of their subsidiaries at will, but are also strong enough to effectively influence the existing legal systems.

Another disturbing example is the North American Free Trade Agreement (NAFTA). Chapter 11 of this document allows private investors to change the local law in court proceedings, with respect to a wide range of issues, such as the environmental impact of a given business. Corporations immediately made use of this option and won some of the lawsuits, resulting in millions of dollars worth of losses to the states involved (Mann & von Moltke, 1999).

Other aspects of globalization are even more dangerous.

First, it concerns the migration of the most skilled and usually best-educated citizens of poorer countries to richer ones. These people contribute with their work to the further development of the North and impoverishment of the South.

Second, large corporations have become supra-national (post-national, and sometimes even anti-national), which has undermined the role of traditional states. If solutions and strategies (at the political and legal levels), introduced by a given country or group of countries (even within the EU), happen to be disadvantageous for these corporations, they will transfer this part of their activities to other countries where such operations might be allowed. Moreover, as many economists point out, corporations are also able to block the development of almost any company not yet owned by them (Ikerd, 2005).

A new trend is the involvement of corporations in military operations and restoration activities after natural disasters. In her book "The Shock Doctrine, The Rise of Disaster Capitalism" (2007) Naomi Klein illustrates that such policies may be very lucrative for large companies. The title of the book refers to the military doctrine adopted by the USA in 2003 during the invasion on Iraq. Klein asserts that the aim of the military operation 'Shock and Awe' was to induce fear and feelings of danger and to wreak havoc on such a scale as to exceed the ability of people, the authorities, and society as a whole to understand it. Also natural phenomena may induce similar effects (Klein, 2007). According to the author, this induced fear is used to rob the attacked countries of their natural resources (such as oil in the case of Iraq), and the process of subsequent restoration is conducted—in order to maximize profits—only with the participation of companies controlled by the 'invaders' and without regard for local people.

The Baghdad Children's Hospital is an example of this mechanism. When Klein visited it in 2004, the facility, built by American developers, failed to meet basic sanitary standards—sewage was draining along corridors and all the toilets were out of order. What is more, although the corporations had received funds for the reconstruction of 142 health centers all over Iraq, only six of them had actually been completed (Klein, 2007).

A similar policy is used in the case of natural disasters, such as Hurricane Katrina, which struck New Orleans in 2005. Klein writes that the money which was meant as support for flood victims was partly used for the liquidation of public schools and their replacement with private ones. There were 123 public schools in New Orleans before Katrina; after the 'restoration' the number has shrunk to four (Klein, 2007).

It is thus not surprising to hear voices warning that the strategy of corporations is to subordinate countries and societies to themselves (or, more frankly, to the liberal global market).

Such actions lead to social and political Darwinism (Kozak, 2001) and mark the return of colonialism, but under new 'banners'. One should note that in Darwin's original theory each species struggles for survival, but also pursues its own benefits, such as individual skills. The development of skills in the course of evolution has resulted in an enormous diversity of higher forms of life. Therefore, adapting 'the struggle for existence' to the principles of economics should not lead to granting the stronger individuals the right to exploit the weaker ones. Darwin described the emergence of diversity, but the large corporations destroy it (Weizsäcker et al., 1999).

The modern social Darwinism assumes a specific kind of selection. In view of the shrinking supplies of natural resources, it rejects the principle of 'common good' which implies the relief of pressure on the environment, especially on the part of the North which is consuming most resources. The North engages all of its scientific and technological strength, but does this only for its own use, leaving the rest of the world to fend for itself. As it was expressed in the report of the Hammarskjold Foundation for the United Nations, the rich are floating in a lifeboat surrounded by the poor swimming in the water. Because the boat is already full, taking the poor on board would cause it to sink with all passengers. So let the poor drown (Michnowski, 2007). In the words of Noam Chomsky, within the framework of this specific 'socialism for the rich' (Gawor, 2006) the literature indicates that the '20:80 principle' applies. This means that 20% of mankind should enjoy a high standard of living and 'have a supply of food', while the remaining 80% should be deprived of these (Martin & Schumann, 1999; Michnowski, 2007). Taking into account the present distribution of goods and the level of resource consumption by the rich, it seems that the '20:80 principle' is already being enforced. However, a new factor has appeared which might significantly change the situation. The '20:80 principle' does not take into account very populous countries, such as China, which recently have been growing rapidly, and whose environmental impacts are just as considerable as that of the North.

Nevertheless, the scientific and technological factors remain in force. It is already possible to use them to shape the environment almost at will. Obviously, this potential is not commonly accessible, because of the rich countries' reluctance to share their achievements (which have economic value).

These constitute the 'internal barriers' the abolition of which was regarded as a 'moral imperative' as early as 1977 in one of the reports prepared for the Club of Rome (Laszlo, 1977).

Thus, Zdzislawa Piatek is right in stating that contemporary globalization deepens the divide between the rich and the poor. It separates instead of uniting (Piatek, 2006).

This aspect is one of the primary motivations behind the currently developing globalization-contesting movements. Within this trend, two basic standpoints are distinguished—anti-globalism and alter-globalism. Both movements are supported by environmentalists, human rights advocates, anarchists, and local community advocates, as well as the supporters of strong national states.

Anti-globalism is a spontaneous movement of protest against globalization, without an institutionalized form, or even uniformity. It is a wide stream of various social, political, and civic initiatives (Gawor, 2008). According to anti-globalists, the landmark event of the movement was the protest against the World Trade Organization conference in Seattle in 1999. The number of protesters was estimated at between 50,000 and 100,000 (Globalissues.org, 2008). In the period that followed, almost every economic summit was accompanied by such protests, with these sometimes taking a violent course. Supporters of this trend emphasize not only the dangers stemming from globalization changes to the economy and environment, but also the importance of social issues (e.g. structural unemployment). Moreover, they believe that one of the major threats of globalization is a mass culture ousting local cultures, resulting in a homogeneous pop-cultural model, figuratively termed 'McDonaldisation'. In view of that, it is not coincidental that one of the major publications about globalization, written by already mentioned Naomi Klein, is entitled "No Logo" (Klein, 2000), which is a protest against popular trademarks owned by large corporations.

Anti-globalism stands up against globalization as such, offering virtually nothing in its place. It is different in the case of alter-globalism. Supporters of this movement notice both the negative and positive impacts of the globalization processes and act strongly against the former. Globalization should serve man and oppose his reduction to an economic indicator. Also, it should not lead to cultural uniformity. In this context, the famous environmental slogan was adopted, 'Think globally, act locally'. It is understood as 'glocalism' (a combination of two seemingly opposite terms, globalism and localism), which means using globalization processes so as to avoid losing the local color (Gawor, 2008).

With reference to the above, one might ask the question: is sustainable development an alternative to globalization?

It is suggested that globalization and sustainable development are two sides of the same coin. According to Duncan French, when globalization reorganizes the world, sustainable development indicates the threats entailed by this new order and which result from the earlier history of mankind (French, 2002).

Let us remember that globalization may be based either on the currently prevailing egoistic axiology, or on eco-humanistic (inclusive) axiology, referring to the common good.

186

Egoistic globalization, in the words of Pope Benedict XVI, seemingly "is not a synonym for world order, but just the opposite. Conflicts generated by the pursuit of economic supremacy and ensuring access to the supply of energy, water, and natural resources are frustrating the efforts of those trying to bring more justice and solidarity to the world. (…) It has become obvious that only by adopting a moderate lifestyle, accompanied by serious endeavors for an equal distribution of wealth, would we be able to achieve fair and sustainable development. In order to do this, we need people who have a great hope and great courage" (Benedict XVI, 2008). This is a continuation of the thought of Pope John Paul II who, in his 1987 encyclical "Sollicitudo rei socialis", frequently addressed the threats of globalization (particularly with regard to the rich exploiting the poor). We might quote his statement of 2001 when he said that, "Globalization, *a priori*, is neither good or bad. It will be what people make of it" (John Paul II, 2001).

But will the path of inclusive globalization—for the common good—be chosen by mankind?

Even now it has been supported in CIA reports. In the 2000 study "Global Trends 2015: A Dialogue About the Future with Non-government Experts", published at the end of Bill Clinton's presidential term of office, inclusive globalization was compared to the currently prevailing egoistic globalization which was defined as downright destructive. It was acknowledged that it is only beneficial to the rich part of the world's society and thus constitutes a considerable threat to mankind as a whole. These conclusions have not been challenged to this day.

There is still much to do not only on the global, but also on the local level where we function directly, and where jobs are created or lost.

This issue is also present in the discussions about sustainable development. Should this development include the growth of the labor market or, maybe, create other activities equal to work in a better way, which also can provide people with basic material living conditions? How can we counteract the overwhelming priority of production efficiency? Is it possible that everyday work could be a motivation to integrate and cooperate, and not to compete? Can this be reconciled with environmental protection?

These questions may be answered positively in view of the currently emerging concepts of regional, sustainable economies as a counterpoint to egoistic globalization. One such example comes from Germany.

According to Reinhard Stransfeld (Stransfeld, 2001), the regional level (so-called 'regional economics') is an opportunity to solve many problems, namely:

- Overcoming the employment crisis.
- Resistance to the disruptions of speculative capital flows.
- Development of a diversified labor market providing opportunities for people with different abilities and skills.
- Creating fields of activity for those who cannot find their place in the global information society.
- Transparency of the life and action space, achieving cultural identity and responsibility.
- Rationalization of the regional circulation of resources and emphasis on a sustainable economy.
- Support for poor countries from rich societies.

Studies of this kind are worthy of special attention. The slogan 'Think globally, act locally' is still up-to-date. The global civilization change, postulated by many modern development strategies, begins with small local steps. It is also the right beginning for a new era—the era of sustainable development.

3 BREAKTHROUGHS IN HUMAN HISTORY

The new vision of sustainable development, discussed in this book, presents it as an interdisciplinary program touching all human activities (global, regional, local), in which we can distinguish three levels and nine planes (ethical, ecological, social, economic, technical, legal and political). Such a wide scope of issues—together with the equally wide-ranging changes proposed for the different planes and the strategies already adopted—allows us to formulate the following claim: should sustainable development actually ever be implemented, it will introduce a new order as revolutionary as the breakthroughs in human history that are conventionally termed 'Revolutions' (Pawłowski, 2008).

The key point of reference here is the change involving not only the relations between man and nature but also person-to-person relations.

From the point of view of biology, it is assumed that modern hominids (a family of primates including man and his ancestors) evolved from beings similar to anthropoid apes inhabiting Africa in humid equatorial forests in a period from 8 to 4 million years ago. It is assumed that the initial species was the *homo habilis*. The *homo sapiens* is said to have appeared in the Lower Palaeolithic (300,000–400,000 years BC), and the form biologically identical to modern man— *homo sapiens sapiens*—35,000–40,000 BC (Bielicki, 1995; Halaczek, 1991).

It is possible to delineate several main stages in human history (Pawłowski, 1999, 2002):

- hunter-gatherer culture,
- agriculture,
- industry and technology.

Some authors seek to further simplify this division.

Philippe Saint Marc (1971) refers only to the Agricultural Age (extending it to the end of the 18th century) and the Industrial Age (up to the middle of the 20th century).

A similar view was adopted by Alvin Toffler, who postulates two stages (waves) in human history: the agrarian revolution, and the creation of the industrial civilization (Toffler 1980, 1995).

Interestingly, both authors choose to roam into the future. Saint Marc envisages the need for his division to take account of nature's age, which Toffler in the broader dimension of his original vision of ecological civilization will point at the third wave.

We may also refer to yet another division proposed by Christian Thomson, which is based on the criterion of the tools used by prehistoric man (Davies, 1998; Monhemius, 2000).

It distinguishes:

- the Stone Age (up to 4000 BC),
- the Copper Age (after 4000 BC),

- the Bronze Age (from 2500 BC),
- the Iron Age (from 1000 BC).

Different view was proposed by Neil Postman (Postman, 1992), indicating what follows:

- Tool-using cultures.
  This term refers to a situation in which the tools are integrated with culture and technology is not able to subject people to its needs. This usually refers to theocratic cultures, i.e. such where theology provides the rationale for human actions and thoughts.
- Technocracy.
  This term refers to a situation where the tools play a central part in the creation of human culture. Tools attack culture, they even suggest that they are the culture themselves. An ideal technocrat is a man without beliefs and own point of view, instead possessing a number of sellable abilities.
- Technopoly.
  This refers to a totalitarian technocracy, i.e. subordination of all forms of cultural life to the rule of technology and engineering. In Technopol, the technical calculation dominates human assessment in every respect. Postman also notes that if in the Middle Ages people used to believe—regardless of circumstances—in the authority of religion, in Technopol they believe—also regardless of circumstances—in the authority of science. *À propos*, according to Postman, the only Technopol currently existing in the world is the USA.

However, the most frequent references (Malinowski & Strzalko, 1985; Young, 1971; Reichholf, 1992; Campbell, 1995) are made to the hunter/gatherer era, as well as to the Agricultural, Scientific and Industrial Revolutions—see Table 49. I think that the list should be augmented by the Sustainable Development Revolution taking place today.

The hunter-gatherer period coincided with the Upper Palaeolithic. Humans inhabited tropical forest and savannah biomes. They were nomads, and were small in numbers thus exerting a minimal—merely local—impact on the environment, until the mastery of fire afforded more opportunities.

The great change came with the Agricultural Revolution, when the spread of purposeful crop cultivation and animal husbandry necessitated a more settled way of life.

The first traces of this kind of management happened 20,000 years ago in the Nile Valley. The innovation proved to be quite easily abandoned, however, when climatic changes offered better access to food everywhere. The agricultural changes accelerated around 9000 years ago in Asia, and in Europe 4000 years later (Malinowski & Strzalko, 1985).

The new conditions brought about stabilisation of food supply and thus allowed humans to direct some of their intellectual energies to activities other than ensuring mere survival (Curtis & Lamberg-Karlovsky, 1973). As early as in the Neolithic period, the smelting of lead and copper was mastered, and the Sumerians developed a pictographic system of writing. It was also at this time that the first settlements were transformed into city-states (Young, 1971). This was a qualitative change. Before that man had done little more than adapt to environmental conditions. With the creation of ever-larger

Table 49. Key stages in mankind's development (Author's own work).

Name of stage	Time period referred to
Hunter-gatherer era	Upper Palaeolithic.
The Agricultural Revolution	Starting c. 9000 years ago in Asia, some 4000 years later in Europe.
The Scientific Revolution	1543—symbolic beginning marked by the publication of Copernicus's "On the Revolutions of the Heavenly Spheres". 1687—development, with the appearance of Newton's "Principia Mathematica".
The Industrial Revolution	1769—symbolic milestone: Watt's improvements to the steam engine. Further stage (1860–1914) starting with the use of oil (in the internal combustion engine) and electricity.
The Sustainable Development Revolution	Three watershed dates: 1969 U'Thant's Report. 1987 definition of sustainable development adopted by the UN. 1992 UN Earth Summit in Rio de Janeiro. 2002 IN Earth Summit in Johannesburg.

settlements he set in train a process of changing the environment to suit his needs. Even if the environmental consequences remained limited at that stage, from then on man's far-reaching interference in the structure of ecosystems never ceased.

A further factor conditioning mankind's advancement was the founding of science and the accompanying development of technological capabilities.

A particular role was played here by the groundbreaking mediaeval discoveries. These included the compass that made possible long-distance sea voyages and brought new geographical discoveries which pushed back the frontiers of human knowledge.

Of no less significance, according to Francis Bacon, was the invention of printing and gunpowder (Bacon, 1620).

Neil Postman stresses the special significance of three inventions (Postman, 1992), i.e.:

- the mechanical clock—ushering in a new concept of time, and consequently the development of Mechanism;
- the telescope—observation of the sky through it led to questioning the belief that the Earth was the centre of the universe;
- the movable-type printing press (reducing the importance of the oral tradition and allowing dissemination of knowledge previously reserved for the privileged few).[15]

---

[15]Before the year 1500 in Europe (with 70 million inhabitants) 20 million books were already printed (Braudel, 1985; van Doren, 1997). The use of the printing press would have never brought such effects if it was not for cheap paper produced from fabric, rags, and old clothes (a technology invented probably in China in the 2nd century). In Europe, the manufacture of large amounts of such paper began in the 14th century. It was the time of the Plague ('the Black Death') which took the lives of about 25–40 million people. One of the unexpected consequences of this situation was a huge excess of clothes and rags, which were subsequently used for the mass production of paper. Its price dropped very low, which contributed to the cheapening of books (Davies, 1998).

The above discoveries are highlighted as factors paving the way for the next great revolution in human history—the Scientific Revolution (Davies, 1998). This marked a shift from Christian to mechanistic natural philosophy. In other words (Brown, 1986, Crombie, 1994, Zieba, 1996, Hajduk, 1996), it was a transition from essentialism (i.e. a search for the essence of nature) to existentialism (i.e. analysis not of the essence, but of the phenomena or manifestations of nature).

The symbolic opening of the Scientific Revolution was marked by the publication in 1543 of "De Revolutionibus Orbium Coelestium" (On the Revolutions of the Celestial Spheres) by the astronomer (mathematician, lawyer and economist) Nicholas Copernicus, and was augmented by the 1687 publication of "Philosphiae Naturalis Principia Mathematica" (Mathematical Principles of Natural Philosophy) by Isaac Newton—the founder of differential and integral calculus, and the laws of motion (Kaminski, 1992).

Also worthy of mention among the most outstanding figures behind the new scientific paradigm are (Kaminski, 1992; Brown, 1986) Galileo Galilei (1564–1642, astronomer and founder of modern physics), Francis Bacon (1561–1626, propagator of the experimental method and originator of inductive reasoning), and Rene Descartes (1596–1650, philosopher regarded as the founder of modern intellectual culture).

The Scientific Revolution meant not just the alteration of a few key theses, but in fact the upheaval of the entire scientific paradigm. The new paradigm that emerged was founded upon reductionism and a mathematical view of nature (Zieba, 1992). Reductionism demanded that nature be studied by dividing the overall complex picture into its simpler component parts, which might then be described with the aid of mathematics (Melsen, 1963).

The most visible consequence of that new stage in the development of science was the rapid expansion of the scope of human knowledge. Until then, all of the said knowledge had in principle been available to anyone ambitious enough to embrace it, while its practical implications were also on a similar scale, ensuring that—in theory at least—anyone could substitute for anyone else. Subsequently, as specialization began to set in, this rapidly became less and less possible, or even imaginable.

It is worth adding that the Scientific Revolution coincided with a period of considerable economic growth that earned it the title of 'the Golden Age'. This was connected with large-scale cultivation of cereals in Europe, and hence with wider and fuller access to good nutrition (Sörlin, 1997). What the landlords of the time were to learn to their pain was that continual husbandry of the same land without the use of fertiliser (not practised at the time) led to soil degradation. Crop yields did indeed begin to decrease, and famines became ever more common. Agricultural output collapsed, which lead to a series of armed conflicts, the first of which broke in 1650 (Sörlin, 1997). The grain trade did not vanish, but never was a big as in 'the Golden Age'. Lack of environmental knowledge led to soil degradation with dire consequences for the society.

Despite their huge scale, these negative phenomena did not hold back the next breakthrough—the Industrial Revolution. The symbol of it was the 1769 patent for an improved steam engine that Scotsman James Watt had begun working on in 1765.[16]

---

[16]It is true, that steam engines were already present in Alexandria in the 1st century BC, however, they were in the nature of curiosities, having no practical application (Encyclopaedia Britannica, 2000).

Watt's work took place in England, and contributed to the rapid development of the British textile industry. At the same time, because wood was used as fuel to produce the steam that powered the looms, demand for firewood soared and most forests were cut down.

On a broader scale, the Industrial Revolution brought about the emergence of factories and the factory system of mass production. Subordinated to it were diverse human activities, all leading to the development of the technological world as we now know it. Of particular importance here was Benjamin Franklin's work on electricity around 1750, as well as the breakthroughs made by Alessandro Volta in 1800 which made practical use of electricity feasible (Doren, 1997). Another source of energy was needed, and that proved to be coal. Initially used for heating dwellings and for cooking, its suitability for powering industrial machinery only came to be appreciated later. Coal energy made mass production possible, but brought the side effect of severe degradation of the environment.

Today's environmental problems are also largely connected with the burning of fossil fuels. But is this process still part of the Industrial Revolution, or are we in fact living in a post-industrial civilization?[17]

4 SUSTAINABLE DEVELOPMENT AS A CIVILIZATIONAL REVOLUTION

The current phase to mankind's development has not yet been unequivocally defined. Industry will certainly continue to play an important role in shaping our civilization, but a series of new phenomena have also emerged. Are these changes bringing along with them yet another revolution?

Some authors say yes, suggesting that we are now dealing with a Modernisation Revolution, whereby the agricultural society living in the countryside is being transformed into an urban industrialised society. This process would not be possible had it not been preceded by the Scientific and Industrial Revolutions. Among other effects, they ushered in the development of a new type of contemporary urban infrastructure (clean running water supply, sewerage, waste collection, goods and passenger transport, food supply, the labour market and health service), thus ensuring the safe functioning of hundreds of thousands of people at one location.

Does mankind's move from the countryside to towns warrant the use of the term 'Revolution'? The negative impact of human activities has certainly affected urban environments more than the rural, to the extent that the former's spread has further increased human pressure. Nonetheless, it is not changing the mode of shaping the relationship between human beings and nature, so it is just another stage of the Industrial Revolution.

There are also opinions that we are now in the middle of an Information Technology Revolution linked with the widespread use of the Internet (Haliniak,

---

[17] That said, there is no consensus as to how many civilizations have thus far taken shape on Earth. A.J. Toynbee for example distinguishes 21 significant examples, which is to say the Egyptian, Andean, Chinese, Minoan, Sumerian, Mayan, Yucatan, Mexican, Hittite, Near Eastern, Babylonian, Persian, Arab, New Chinese, Korean-Japanese, Indian, Hindu, Greek, Byzantine-Orthodox, Russian Orthodox and Western European (Toynbee, 1948).

2004; Michnowski, 2003). Indeed, the Internet is an exceptional platform allowing access to and dissemination of crucial information, contributing to the development of an information society (Michnowski, 2003, 2006).

At the same time, the technology itself seems to be heading down a blind alley. As of 2007, as many as 95% of all e-mail messages received by users were spam, i.e. unsolicited material containing brazen advertising; the respective figure for 2001 had been just 5% (PC World Computer, 2008). What is more, the authors of spam messages are hiding behind well-known institutions and websites. This falsification of identity is a widespread phenomenon, with many Internet users receiving mail (usually advertisements) allegedly sent from their own addresses but actually authored by someone else. The next step is web-page forgery, e.g. of popular banks. The false copies look identical to the originals, differing only in the address. If we were tricked into submitting such data as our PIN codes, the fraudsters are able to empty our accounts.

The Internet has not changed man's attitude to nature, while the ever more sophisticated channels of communication between people have done nothing to limit the pressure they exert on the environment. It may therefore be just a tool put to work for a revolution of a more general scope, just as Watt's steam engine was merely a symbol of the Industrial Revolution.

What in that case may bring about the desired change?

Perhaps the current debate on sustainable development does deserve to be called a revolution in its own right (Pawłowski, 2009, 2010). While it is true that this kind of development has not been introduced yet, many recent political, legal and economic initiatives are heading in that direction. Man's current influence on the biosphere is without doubt global in reach, and hence it requires a world-wide sustainable response. The idea of sustainable development emphasises the necessity for basic change in mankind's behaviour in relation to the environment as a 'toll' to stop further destruction and thus save the planet; it is trying to cure the causes not only the consequences of the present problems—it is much more than a Modernisation Revolution or Information Technology Revolution.

In the article "Human Progress Towards Equitable Sustainable Development: A Philosophical Exploration", together with Victor Udo we explored the progress of humanity and the challenge of sustainability from the first human to the contemporary using the Cascaded—S Curve. We observed that major historical and teleological human progress tends to occur in five critical stages: the existing level of progress, crisis-breakthrough stage, breakthrough education stage, transformative action stage, and a new level of progress stage—see Table 50.

So, how can stage 5 be achieved and implement a new, sustainable level of progress?

Any attempt at assessment of the Sustainable Development Revolution is impeded by the relatively limited time span available to us. Moreover, it remains hard to say much about the future as, at any moment, factors changing our present vantage point may swing into action. Just as the September 11th terrorist attacks in New York obviously shattered the illusion that today's world can offer security, unexpected environmental disasters may also strike as a result of man's activities.

But equally well we might witness new groundbreaking scientific and technological discoveries opening the way to new and efficient energy sources providing an alternative to the diminishing reserves of fossil fuels.

Table 50. Summary of S-Curve model applications in society (Udo & Pawłowski, 2010).

Phenomena/ stages	Idea diffusion/ Product life cycle	Economic growth	Scientific Revolution	Global Human Social Progress (GHSP)
Stage 1	Innovators/ introduction	Traditional society	Old paradigm	Existing level of progress
Stage 2	Early adopters/ early growth	Pre-conditions for take off	Anomalies	Crisis-breakthrough
Stage 3	Early majority/ late growth	Take off	Crisis	Breakthrough-education
Stage 4	Late majority/ maturity	Rise to Maturity	Extra science	Transformative action
Stage 5	Laggards/ decline	High mass consumption	New paradigm	New level of progress
Application area	Technology/ business	Political economy	Social science	Applied sustainability
Issues addressed	Economic-profit & market share	Equity-wealth re-distribution	Empowerment-political power	Sustainable development revolution, so achieving equitable global sustainable development (EGSD)

Despite the high level of uncertainty about the future, reference also has to be made to the past. It is no secret that a number of today's environmental problems had originated well before the Industrial Revolution got underway. The technical solutions are already available—it is time to introduce them.

We should also think about us more personally. A hint may be found in books about the ecology of man. It concerns the development of the so-called 'thinking tissue', i.e. the human brain. Its mass, in reference to the entire human population, has grown through history from about 300 tonnes at the beginning of human existence, to 270 000 tonnes at the start of the present era, to approximately 7,000,000 tonnes today (Wolanski & Siniarska, 2001). We are becoming more and more powerful, so let us use this power for our own good—by introducing sustainable development.

Conclusions

The concept of sustainable development has become practically a household term nowadays. When I finished writing this book, on 14 March 2011 a Google search for the term 'sustainable development' returned 23,900,000 results. This is a lot, especially bearing in mind that the principle of sustainable development was only introduced slightly more than two decades ago. This principle invokes the rights of the present and future generations and it can be considered on three levels:

– first, referring to the ethical values, constituting the 'foundation' of human activity,
– second, consisting of equally treated social, environmental, and economic issues,
– third, covering the discussion on detailed questions pertaining to law, politics and technology.

Thus, beginning with philosophy we close with detailed issues connected with the selection of pertinent technological solutions.

At their inception, the strategies referring to sustainable development focused mainly on environmental matters. This was probably due to the fact that an important thread in human history was the development of abilities, leading to ever stronger control—and degradation—of nature. The primitive Palaeolithic hunter and nomad has come a long way to become a modern being living in a heated house in which the conditions may be almost freely regulated: both in terms of air composition, and temperature. However, he was not able to grow completely independently of the external environment, even if he seized the power to destroy it. Are we thus doomed to self-destruction? Great extinctions have already occurred in the history of planet Earth—will man be responsible for the next one?

Warnings against self-destruction were already appearing in late 19th century. What is interesting to note, they occurred simultaneously with the introduction of the term 'modernity' to colloquial language (Zgoda, 1988). Its popularity was established in the 20th century which can indeed be described as modern. At the same time, it was a real challenge to mankind. The landmark moments were marked by two world wars. Their consequences affected all realms related to man, including also problematic issues which, in the present study, were referred to as the levels of sustainable development:

• Radical changes occurred with respect to society and ethics. Mankind has proved itself to be capable of exterminating entire nations, as particularly illustrated by the tragedy of the Jewish nation (the Holocaust). However, after that happened, it was able to institute a number of organizations whose aim was to prevent such occurrences in the future. Since the beginning of this initiative a special part was played by the United Nations.

- In terms of environment and ecology, the wars resulted in immense damage. The first one was fought only in Europe and brought about numerous local changes, but the second one was global.
- Not only did the area of military efforts become greater, but also the technical powers increased immensely, suffice to mention the atomic bombs dropped on Hiroshima and Nagasaki.
- As far as the economic aspect is concerned, the support programs established after the war and aimed at reconstruction and development of different world regions had a significant impact. What is important, these programs were not based solely on purely economic operations, but also took into account a wider spectrum of problems, including social issues.
- Within the legal and political level, the wars completely revised the previous order. There appeared two mutually hostile camps of the East and the West which established independent and internal forms of cooperation within the EEC (European Economic Community) and Comecon (Council for Mutual Economic Assistance).

When the principle of sustainable development was formulated in 1987, this division was still present. It was only the transformations in Eastern Europe, initiated by Poland in 1989, that changed the situation. The free market, until then a domain of the North, became a global paradigm. However, it failed to solve the problems of mankind, even the most fundamental ones. While successive development programs are established in the name of fairness and equal opportunities, the gap between the rich and poor countries is still widening. The situation is not changed even by desperate initiatives of the poor countries, such as the establishment of the G-15 group,[18] intended to counterbalance the 'wealthy' G-8.

Moreover, the principal environmental problems of the modern world came into existence largely due to various practices of the highly-developed countries of the North. Also now, due to the wealth they own, the most is expected of them. Unfortunately, many of the operations of the Northern countries remain unsustainable. Their decisions about further development will have global significance, due to their economic and political strength, and the imposed model of behaviour.

Will the North change its face? I believe it is the consistent implementation of sustainable development that provides a chance for real improvement. True, this type of development has not been introduced yet, but many current political, legal, and economic initiatives are heading in this very direction. If they manage to succeed in the long run, we will be even able to speak about a sustainable development revolution, comparable to the scientific or industrial revolutions. This is because it would imply a change in the attitude of man to his environment, and in his inter-personal relations.

The postulated revolution is thus an optimistic vision of human development, not only assuming that the catastrophe can be avoided, but also showing completely the new possibilities mankind could introduce. They are described in the adopted programs of sustainable development. However, the proposed

---

[18]Currently composed of the following 18 developing countries: Algeria, Argentina, Brazil, Chile, Egypt, India, Indonesia, Iran, Jamaica, Kenya, Malaysia, Mexico, Nigeria, Peru, Senegal, Sri Lanka, Venezuela and Zimbabwe.

recommendations require a broader motivation than only through data, tables, and graphs. After all, it is completely different when people act in certain ways because the law tells them so, than when they are deeply convinced about it and it complies with their personal value system. The latter element is one of the issues considered within the field of eco-philosophy, or more precisely, environmental ethics. Therefore, I think that, without an ethical (or eco-philosophical) foundation, the sustainable development revolution cannot succeed. Yet, philosophers and ethics do not participate in studies on the strategies of sustainable development; it is therefore even more necessary to emphasise and deepen the philosophical elements within all levels of sustainable development. One of the ways to achieve this aim might be the idea of philosophical audit described herein. This new research field still requires the development of clear methodological foundations, but it already seems to be an interesting option.

Discussion must also be continued within the framework of the 'basic' ethical level of sustainable development. From the formal point of view, let us remind ourselves that the principle of sustainable development is anthropocentric because it concerns the human living environment and its quality. It is also very distinctly future-oriented (the well-being of future generations), thus related to the purposeful activity of man, being exactly the reference to the future and its dreams and hopes. The future is unknown, but each choice we make (both right and wrong) contributes more or less to its approximation and determination.

There is still the question of diversity of eco-philosophical standpoints. Regardless of the argumentation being religious or lay, anthropocentric or not, it is important to protect the environment. From the point of view of nature, our motives do not matter, what matters is specific actions taken on the basis of the motives. In this context I represent the 'moderate anthropocentric view'. I think that nothing can make man act more than his own interest. If environment is indispensable to man—it should be protected for this very reason.

In this context it is a great challenge, taking into account the increasingly clearer threats related to global consequences of human activity in the environment. The recent years have admittedly brought many environment-friendly technologies, but very often they are too expensive, especially for the poor countries of the South. Not being able—at least yet—to introduce them all, choices must be made. What needs our attention first?

There are certain new tools which may be of use, such as backcasting which employs comparative futurology to help make the right and sustainable choices today.

Also, as far as detailed solutions are concerned, one might point to the concept of 'life boats'. This is a postulate to protect these ecosystems first whose natural wealth is most beneficial to mankind. They are areas with a high level of biodiversity but usually situated in regions of poverty, where people's survival is virtually completely dependent on nature. In such conditions the environmental pressure is intensified, degradation processes are accelerated, and, as a result, nature's ability to meet the needs of local communities decreases. However, if the basic needs of the local communities were satisfied, the pressure on the environment would immediately lessen. According to P. Kareiva and M. Markier, the relationship between saving endangered species and human welfare, though not always apparent at first glance, is undisputable. A specific example of such a dependence is the circumstances which led to introducing three species of vultures to the list of critically endangered species: the Indian vulture, the slender-billed vulture, and the Indian white-rumped vulture. Their populations were rapidly decimated in the 1990s. The cause turned out to be a result of the common use of an anti-inflammatory drug (diclofenac), relatively harmless to people but lethally dangerous to birds. The vultures used to eat thousands of cow carcasses which were intentionally left for them according to Indian tradition. But when the population of birds diminished, there was too much rotten meat, which led to the spread of dangerous anthrax bacteria. Moreover, the meat became an easy prey for stray dogs, which made their population grow, and increased the threat of rabies (Kereiva & Marvier, 2007).

In view of this, nature conservation should go far beyond the proposals of establishing more areas protected by law, becoming a real and important component of general development processes concerning everyone. Such a strong link between the issue of ecology and the development of mankind is in accordance with the principle of sustainable development. What is important, its benefits will be enjoyed by both—people and the environment. For example, we might invoke the problem of the lack of access to drinking water which concerns about 1 billion people globally and claims 3 million victims each year. The situation may be improved not only with technical means but also by cooperating with nature. In this case it would be a priority to protect forests and boggy areas which function as natural 'sewage treatment plants' filtering out the pollutants.

Will the future world be sustainable then? Can *homo sapiens* become *homo sustinens* (Sienhuner, 2000)? There is no simple answer to these questions. However, irrespective of the degree to which the current situation fulfils our expectations (or fails them), a lot has been achieved since the publication of the report on "Our Common Future", as we are trying to show in our journal "Problemy Ekorozwoju/Problems of Sustainable Development" (ecodevelopment. pollub.pl). The sustainable development revolution has begun.

References

Ackerman, F. 1997. *Why Do We Recycle? Markets, Values and Public Policy*. Washington: Island Press.

Ackerman, F. 2001. Materials, Energy and Climate Change. In J.M. Harris, T.A. Wise, K.P. Gallagher & N.R. Goodwin (eds), *A Survey of Sustainable Development. Social and Economic Dimensions. Frontier Issues in Economic Thought*: 198–199. Washington, Covelo, London: Island Press.

Adamek Z. 2001. *Elementy wiedzy o kulturze*. Tarnow: Biblos.

Adreasson-Gren, I.M., Michanek G. & Ebbesson J. 1992. *Economy and Law – Environmental Protection in the Baltic Region*. Uppsala: Uppsala Publishing House.

Agenda 21. The First Five Years. Implementation of Agenda 21 in the European Community. 1979. Luxembourg: Office for Official Publications of the European Communities.

Agenda *21. The United Nations Programme of Action from Rio. The Final Text of Agreements Negotiated by Governments at the United Nations Conference on Environment and Development (UNCED) 3–14 June 1992*. 1992. Rio de Janeiro: UN.

Agger, B.P. 1991. Ecological Consequences of Current Land Use Changes in Denmark and Some Perspectives for Planning and Management. In *Proceedings of European Seminar on Practical Landscape Ecology*: 4–5. IALE: Rockslide.

Albinska, E. 2005. *Czlowiek w srodowisku spolecznym i przyrodniczym*. Lublin: KUL.

Albrecht, J. 2006. Green Tax Reforms for Industrial Transformation: Overcoming Institutional Inertia with Consumption Taxes. *Sustainable Development* 14: 302–303.

The Alcali Act. 1863. http://www.muellerscience.com/WIRTSCHAFT/Umwelt/ Umweltschutz_ seit_ 1200.htm [access 1.12.2010].

Aleksandrowicz, J. 1978. *Sumienie ekologiczne*. Warsaw: WP.

Al-Hadid, W.A. 2002. *Environment and Sustainable Development: Implementation of Agenda 21 and the Programme for the Further Implementation of Agenda 21*. New York: UN.

Allenby, B. 1992. Achieving Sustainable Development Through Industrial Ecology. *International Environmental Affairs* 4 (1): 56–68.

America's Animal Factories: How States Fail to Prevent Pollution from Livestock Waste. 1988. Washington: Natural Resources Defence Council, Clean Water Council.

Amyo, E. 1996. Overwiew of Agriculture in Eastern Europe. In R. Jones, M. Summers & E. Mayo (eds), *Sustainable Agriculture, Economic Alternatives for Eastern Europe* 3: 4–11. London: The New Economics Foundation.

Anderson, A.A. 1996. Uniwersalna sprawiedliwosc a kryzys ekologiczny. In J. Kuczynski (ed.), *Ziemia naszym domem*: 101–112. Warsaw: Scholar.

Andersson, M. 1997. *From Intention to Action. Implementing Sustainable Development*. Uppsala: Uppsala Publishing House.

Andersson, H., Berg, P.G. & Ryden, L. 1997. *Community Development. Approaches to Sustainable Habitation*. Uppsala: Uppsala Publishing House.

Andreasson-Gren, I.M., Michanek, G. & Ebbesson, J. 1992. *Economy and Law. Environmental Protection in the Baltic Region*. Uppsala: Uppsala Publishing House.

Anielak, A.M. 2007. *Polska inzynieria srodowiska – informator*. Lublin: KIS.

Annan, A. 1997. *Renewing the United Nations: A Programme for Reform*. New York: UN.

Annan, K.A. 2000. *We the Peoples'. The Role of United Nations in the 21st Century.* New York: UN.

Armstrong, S.J. & Botzler, R.G. 1993. *Environmental Ethics. Divergence & Convergence.* New York: McGraw-Hill Inc.

1986. *The Assisi Declaration. Messages on Man and Nature from Buddhism, Christianity, Hinduism, Islam and Judaism.* Basilica di Francesco. Assisi: WWF.

Atkinson, G., Dietz, S. & Neumayer, E. 2007. *Handbook of Sustainable Development.* Bodmin: MPG Books.

Ayres, R.U. 2004. On The Life Cycle Metaphor: Where Ecology and Economics Diverge. *Ecological Economics* 48: 425–438.

Bacon, F. 1620. *Novum Organum.* http://www.constitution.org/bacon/nov_org.htm [8.08.2011].

Bachelet, M. 1999. Ingerencja ekologiczna. In J.C. Masclet (ed.), *Wspolnota Europejska a srodowisko naturalne. Konferencja w Angers*: 273–274. Lublin: KUL.

Backlund, P., Holmbom, B. & Leppakoski, E. 1992. *Industrial Emissions and Toxic Pollutants.* Uppsala: Uppsala Publishing House.

Baez, A.V., Knamiller, G.W. & Smyth, J.C. 1987. *The Environment, Science and Technology Education and Future Human Needs.* Oxford, New York, Bejing, Frankfurt, Sao Paulo, Sydney, Tokyo, Toronto: Pergamon Press.

Bajda, J. 1999. Grzech ekologiczny. In J.M. Dolega & J.W. Czartoszewski (eds), *Ochrona srodowiska w filozofii i teologii*: 222–242. Warsaw: ATK.

Baker, S. 2000. The European Union: Integration, Competition, Growth – and Sustainability. In W.M. Lafferty & J. Meadowcroft (eds), *Implementing Sustainable Development. Strategies and Initiatives in High Consumption Societies*: 303–336. New York: Oxford University Press.

Baltscheffsky, S. 1992. Financing the Rio Agreements. Tight Purse Strings in the Rich World. *Enviro* 14: 22–23.

Baltscheffsky, S. 1992. Swedish Economist Carl Folke: Ecologize the Economy! *Enviro* 14: 24–27.

Banka, A. 1993. Przemiana molochow, miedzy psychologia samospelniajacego sie proroctwa katastrofy, a nadzieja na postep. In S. Kyc (ed.), *Kryzys idei postepu – wymiar ekologiczny*: 123–141. Lublin: Politechnika Lubelska.

Banka, J. 1996. Eutyftonika – zycie psychiczne czlowieka w systemach technicznych. In S. Zieba & A. Pawłowski (eds), *Technika szansa czy zagrozeniem*: 7–16. Lublin: Politechnika Lubelska.

Barber, B.R. 1995. *Jihad vs. McWorld: Terrorism's Challenge to Democracy.* New York: Ballantine Books.

Baudot, B. & Moomaw, W.R. 1999. *People and Their Planet: Searching for Balance.* London: Macmillan.

Benedict, XVI. 2008. *Statement delivered in the Vatican on 6th January 2008 during the Epiphany Mass.* Vatican: Ignatius.

Benedict, XVI. 2009. *Caritas In veritate.* Rome: Igantius.

Berdo. J. 2006. *Zrownowazony rozwoj. W strone zycia w harmonii z przyroda.* Sopot: Warth Conservation.

Bergh, V. den & Verbuggen, J. & H. 1999. Spatial Sustainability, Trade and Indicators: An Evaluation of the Ecological Footprint. *Ecological Economics* 1(29): 61–72.

Bergström, G.W. 1992. *Environmental Policy & Cooperation in the Baltic Region.* Uppsala: Uppsala Publishing House.

Bhalla, A.S. 1998. *Globalization, Growth and Maginalization.* New York: St Martin's Press.

Bhaskar, V. & Glyn, A. 1995. *The North, The South and the Environment.* New York: St. Martin's Press.

Bialowas, T. 2005. Dynamika gospodarcza i wzrost znaczenia Chin w gospodarce swiatowej. In B. Mucha-Leszko (ed.), *Wspolczesna gospodarka swiatowa, glowne centra gospodarcze*: 301–319. Lublin: UMCS.

Biderman, A.W. 2002. *Poradnik szkoly dla ekorozwoju*. Cracow: PdS.

Bielecki, J. 1986. *Wybrane zagadnienia psychologii*. Warsaw: ATK.

Bielicki, T. 1995. O organizacji spolecznej i ekologii hominidow plejstocenskich. In B. Kuznicka (ed.), *Ekologia czlowieka. Historia i wspolczesnosc*. Warsaw: IHN PAN.

Birnbacher, D. 1995. *Responsibility for Future Genenerations – Scope and Limits*. Warsaw: UKSW.

Bithas, K.P. & Christofakis, M. 2006. Environmetally Sustainable Cities. Critical Review and Operational Conditions. *Sustainable Development* 14177–189.

Blazejowski, J. 2007. The Scientific, Educational, Economic and Social Meanings of Sustainable Development. In L. Pawłowski, M.R. Dudzinska & A. Pawłowski (eds), *Environmental Engineering*: 11–12. London: Taylor & Francis Group.

Bloch, E. 1959. *Das Prinzip Hoffnung*. Verlag: Frankfurt.

Boc, J., Nowacki, K. & Samborska-Boc, E. 2005. *Ochrona srodowiska*. Wroclaw: Kolonia.

Bochenski, J. 1994. *Sto zabobonow*. Cracow: Dajwor.

Bochniarz, Z. 1991. Overview of the Polish Environmental System, Deficiencies and Constrains. In Z. Bochniarz & R. Bolan (eds), *Designing Institutions for Sustainable Development: a New Challenge for Poland*: 19–34. Minneapolis, Bialystok: BTU.

Bodin, B. & Ebbersten, S. 1997. *Food and Fibres. Sustainable Agriculture, Forestry and Fishery*. Uppsala: Uppsala Publishing House.

Bonenberg, M. 1992. *Etyka srodowiskowa. Zalozenia i kierunki*. Cracow: UJ.

Boothroyd, P. 1995. Policy Assessment. In F. Vanclay & D.A. Bronstein (eds), *Environmental and Social Impact Assessment*: 83–126. Chichester, New York, Brisbane, Toronto, Singapore: Wiley & Sons.

Borhulski, S. 2006. Proba ksztaltowania zielonego pierscienia w obszarze aglomeracji lubelskiej. In S. Kozlowski (ed.), *Zywiolowe rozprzestrzenianie sie miast*: 297–298. Bialystok, Lublin, Warsaw: EiS.

Borkowski, J. 1992. Wspolczesne zagrozenia cywilizacyjne. In C. Napiorkowski & B. Koc (eds), *Chronic by przetrwac*: 47–48. Niepokalanow: WOF.

Borkowski, R. 1991. Przed Czarnobylem byl Kysztym. *Aura* 4: 30.

Borkowski, W. 2002. Rodzaje terenowych sciezek dydaktycznych. In J. Lesniewska & J.W. Czartoszewski (eds), *Edukacja w naturze, czyli jak pokochac, poznac, zrozumiec i chronic przyrode*: 49–53. Warsaw: Verbinum.

Borys, T. 1998. Teoretyczne aspekty konstruowania wskaznikow ekorozwoju. In B. Poskrobko (ed.), *Sterowanie ekorozwojem*: 176. Bialystok: PB.

Borys, T. 1999. *Wskazniki ekorozwoju*. Bialystok: EiS.

Borys, T. 2003. Jakosc zycia jako integrujacy rodzaj jakosci. In J. Tomczyk-Tolkacz & A. Ptaszynska (eds), *Jakosc zycia w perspektywie nauk humanistycznych, ekonomicznych i ekologii*: 10. Jelenia Gora: KZJiS.

Borys, T. 2003. Rola zasad w realizacji koncepcji zrownowazonego rozwoju. In T. Borys & Z. Przybyla (eds), *Zarzadzanie jakoscia i srodowiskiem*: 115–129. Jelenia Gora: KZJiS.

Borys, T. & Bryczkowska, M. 2003. Kategoria efektywnosci i jej aspekty srodowiskowe. In T. Borys (ed.), *Zarzadzanie jakoscia i srodowiskiem*: 130–142. Jelenia Gora: KZJiS.

Borys, T. 2005. *Wskazniki zrownowazonego rozwoju*. Warsaw, Bialystok, EiS.

Borys, T. & Rogala, P. (eds). 2002. *Jak opracowac raport srodowiskowy*. Jelenia Gora:FK.

Bosshard, P. 1996. *Lending Credibility: New Mandates and Partnership for the World Bank*. Washington: WWF.

Bowman-Cutter, W., Spero, J., D'Andrea & Tyson, L. 2006. New World, New Deal. A Democratic Approach to Globalization. In N. Haenn & R.R. Wilk (eds), *The Environment in Anthropology. A Reader in Ecology. Culture and Sustainable Living*: 325–326. New York, London: New York University Press.

Boyce, J.K. 1996. Ecological Distribution, Agricultural Trade, Liberalization and In Situ Genetic Diversity. *Journal of Income Distribution* 2: 265–286.

Boyle, G. 1996. *Renewable Energy: Power for Sustainable Future*. Oxford: The Open University and Oxford University Press.

Bratkowski, S. 1991. *Ksztaltowanie i ochrona srodowiska czlowieka*. Warsaw: PWN.

Braudel, F. 1975. *Civilization and Capitalism, 15th-18th Centuries*. New York: HarperCollins.

Breymeyer, A. & Noble, R. (eds). 1996. *Biodiversity Conservation in Transboundary Protected Areas*. Washington: National Academy Press.

Brown, H. 1986. *The Wisdom of Science. Its Relevance to Culture and Religion*. Cambridge: Cambridge University Press.

Brown, P.M. & Cameron, L.D. 2000. What Can Be Done to Reduce Overconsumption? *Ecological Economics* 32: 27–41.

Bruce, J.P. 1996. *Climate Change 1995: Economic and Social Dimensions of Climate Change. Contribution of Working Group III to the Second Assessment Report of the Intergovernmental Panel on Climate Change*. Cambridge: Cambridge University Press.

Bruyn, S.T. 2000. Civil Associations and Toward a Global Civil Economy. In *A Civil Economy. Transforming the Market in the Twenty-First Century*: 290–291. Ann Arbor: University of Michigan Press.

Bryner, G.C. 2000. The United States: „Sorry – Not Our Problem". In W.H.Lafferty & J. Meadowcroft (eds), *Implementing Sustainable Development. Strategies and Initiatives in High Consumption Societies*: 273–302. Oxford: Oxford University Press.

Buckley, R. 1995. Environmental Auditing. In F. Vanclay & D.A. Bronstein (eds), *Environmental and Social Impact Assessment*: 283–301. Chichester, New York, Brisbane, Toronto, Singapour: Wiley & Sons.

Buckinghan, S. & Theonbald, K. 2003. *Local Environmental Sustainability*. Boca Raton, Boston, New York, Washington, Cambridge: CRC Press, Woodhead Publishing Limited.

Burchard-Dziubinska, M. 1998. Rola panstwa we wdrazaniu ekorozwoju. In B. Poskrobko (ed.), *Sterowanie ekorozwojem*: 42–49. Bialystok: Politechnika Bialostocka.

Burdge, R.J. & Vanclay, F. 1995. Social Impact Assessment. In F. Vanclay & D.A. Bronstein (eds), *Environmental and Social Impact Assessment*: 31–65. Chichester, New York, Brisbane, Toronto, Singapour: Wiley & Sons.

Campbell, B. 1995. *Human Ecology: The Story of Our Place in Nature from Prehistory to the Present*. New York: Aldine Transaction.

Carrying for the Earth. A Strategy for Sustainable Living. 1991. IUCN: Gland.

Carson, R. 1962. *Silent Spring*. Boston: Houghton Mifflin Co.

Cello Nieto, C. & Durbin, P.T. 2008. *Sustainable Developlment and Technology*. http://scholar.lib. vt.edu/ journals/SPT/v1n1n2/pdf/durbin.pdf [10.06.2010].

Charmont, J.F. 2006. Interdisciplinarity, a Problematic Solution? In *Sustainable Development, Towards a Sustainable Dialogue Between Science and Policy*: 149–159. Federal Office for Scientific, Technical and Cultural Affairs: Brussels.

Chen, R.J.C. 2006. Islands in Europe: Development of an Island Tourism Multi-Dimensional Model (ITMDM). *Sustainable Development* 14: 104–114.

Chojnicki, Z. 1988. Koncepcja terytorialnego systemu spolecznego. *Przeglad Geograficzny* 60: 498.

Choo, K. 2006. Feeling the Heat. The Growing Debate Over Climate Change Takes on Legal Overtones. *Aba Journal* July: 30–31.

Christen, K. 2005. Turbulence on the Wind Farm. *Environmental Science and Technology* 1 March: 97A–98A.

Christophe-Tchakaloff, M.F. 1999. Unia wobec sektora odpadow. In J.C. Masclet (ed.), *Wspolnota europejska a srodowisko naturalne. Konferencja w Angers*: 246. Lublin: KUL.

Ciazela, H. 2006. Etyka odpowiedzialnosci Hansa Jonasa a trwaly i zrownowazony rozwoj (Imperatyw i dylematy). *Problemy Ekorozwoju/Problems of Sustainable Development* 2(1): 107–114.

Cichy, D., Michajlow, W. & Sandner, H. 1988. *Ochrona i ksztaltowanie srodowiska.* Warsaw: WSiP.

Cichy, D. 2002. Kultura w edukacji srodowiskowej. In J.M. Dolega (ed.), *Podstawy kultury ekologicznej*: 15. Warsaw: UKSW.

Ciolkosz, A. & Bielecka, E. 2005. *Pokrycie terenu w Polsce. Bazy danych CORINE Land Cover.* Warsaw: BMS.

Cities, People and Poverty. 1991. New York: UNDP.

Clark, G. 2007. Evolution of the Global Sustainable Consumption and Production Policy and the United Nations Environment Programme's (UNEP) Supporting Activities. *Journal of Cleaner Production* 15: 492–498.

Clarke, M. & Islam, S.M.N. 2006. National Account Measures and Sustainability Objectives. Present Approaches and Future Prospects. *Sustainable Development* 14: 219–233.

Clarke, T. 2002. Now We're Talking. *Nature* 420: 733–734.

Climate Time Bomb. 1994. Amsterdam: Greenpeace International.

Clunies-Ross, T. & Hildyard, N. 1996. Industrial Agriculture: Unsustainable Face of Raming. In R. Jones, M. Summers & E. Mayo (eds), *Sustainable Agriculture, Economic Alternatives for Eastern Europe* 3: 12–17. London: The New Economics Foundation.

Connolly, P. 1999. Mexico City: Our Common Future? *Environment and Urbanization* 1(11): 53–78.

Constanza, R., d'Arge, R., de Groot, R., Faber, S., Grasso, M., Hannon, B., Limburg, K., Naeem, S., O'Neill, R.V., Paruelo, J., Raskin, R.G., Sutton, P. & Belt, M. van den. 1997. The Value of the World's Ecosystem Services and Natural Capital. *Nature* 387: 253–260.

Constanza, R. & Patten, B.C. 1995. Defining and Predicting Sustainability. *Ecological Economics* 15: 195.

Constanza, R. & Daly, H.E. 2001. Natural Capital and Sustainable Development. In J.M. Harris, T.A. Wise, K.P. Gallagher & N.R. Goodwin (eds), *A Survey of Sustainable Development. Social and Economic Dimensions. Frontier Issues in Economic Thought*: 14. Washington, Covelo, London: Island Press.

Constanza, R. & Farber, S. 2002. Introduction to the Special Issue on the Dynamics and Value of Ecosystem Services. Integrating Economic and Ecological Perspectives. *Ecological Economics* 41: 367–373.

Cooney, C.M. 2006. Sustainable Agriculture Delivers the Crops. *Environmental Science & Technology* 15 February: 1091–1092.

Crombie, A.C. 1994. *Styles of Scientific Thinking in the European Tradition.* London: Duckworth.

Curtis, H., Lamberg-Karlovsky, C.C., Hammond, N. & De Paor, L. 1973. The World Before Civilizations. In J. Roberts (ed.), *Civilization. The Emergence of Man in Society*: 37. Del Mar: CRM Books.

Czaczkowska, I. 1995. *Zagadnienia ekologiczne w pracach i dokumentach Swiatowej Rady Kosciolow. Czlowiek i Przyroda* 2: 86–87.

Czaja S. 2005. Kategoria czasu w ksztaltowaniu zrownowazonego rozwoju. In B. Poskrobko & S. Kozlowski (eds), *Zrownowazony rozwoj – wybrane problemy teoretyczne i implementacja w swietle dokumentow Unii Europejskiej*: 103. Bialystok, Warsaw: K CiS PAN.

Czerniak, J. 2005. Stany Zjednoczone w systemie gospodarki globalnej. In B. Mucha-Leszko (ed.), *Wspolczesna gospodarka swiatowa, glowne centra gospodarcze*: 121. Lublin: UMCS.

Czyz, A. 1992. Przygotowania do Konferencji ONZ „Srodowisko i Rozwoj" w Rio de Janeiro 3–14 czerwca 1992 r. *Srodowisko i Rozwoj* 1: 43.

Daly, H.E. 1996. From Adjustment to Sustainable Development: The Obstacle of Free Trade. In *Beyond Growth: The Economics of Sustainable Development*: 158–167. Boston: Beacon Press.

Danielson, J. 1995. Conference on Flora and Fauna Conservation: Six Measures Could Save Thousands of Species. In *Enviro* 19.

Dasmann, R.F., Milton, J.P. & Freeman, P.H. 1973. *Ecological Principles for Economic Development*. Washington: Wiley & Sons.

Davies, N. 1988. *Europe. A History*. New York: Harper Perennial.

Declaration of the United Nations on the Human Environment.1972. http://www.unep. org/ Documents.multilingual/Default.asp? [30.06.2008].

Dider, J. 1992. *Slownik filozofii*. Katowice: Ksiaznica.

Dobrzanski, G. 1993. Przyrodnicze podstawy ochrony biosfery. In B. Prandecka (ed.), *Interdyscyplinarne podstawy ochrony srodowiska przyrodniczego*: 24–37. Wroclaw, Warsaw, Cracow: Ossolineum.

Dobrzanski, G. 2000. Trwaly rozwoj a teoria ekonomii. In A. Papuzinski (ed.), *Polityka ekologiczna III Rzeczypospolitej*: 382–383. Bydgoszcz: Akademia Bydgoska.

Dobrzanski, G. 2001. Spoleczne aspekty trwalego rozwoju i jego pomiaru. In A. Pawłowski & M.R. Dudzinska (eds), *Zrownowazony rozwoj w polityce i badaniach naukowych*: 164. Lublin: Politechnika Lubelska.

Dolega, J.M. 2006. Ekofilozofia jej otulina. In A. Skowronski (ed.), *Rozmaitosci ekofilozofii*: 45. Olecko: WM.

Dolega, J.M. 2006. Ekofilozofia – nauka XXI wieku. *Problemy Ekorozwoju/Problems of Sustainable Development* 1(1): 17–22.

Dolega, J.M. & Siedlecka-Siwuda, J. 2006. *Zielone Pluca Polski i Europy*. Olecko: WM.

Dolega, J.M. 2006. Miejsce problematyki etycznej i ekonomicznej w ekofilozofii. In W. Tyburski (ed.), *Ekonomia, ekologia, etyka*: 7. Torun: Top Kurier.

Dolega, J.M. 2007. Systemy wartosci w zrownowazonym rozwoju. *Problemy Ekorozwoju/ Problems of Sustainable Development* 2(2): 41–49.

Domanski, R. 2004. *Geografia ekonomiczna, ujecie dynamiczne*. Warsaw: PWN.

Doren, C. van. 1997. *Historia wiedzy od zarania dziejow do dzis*. Warsaw: Al Fine.

Dorste, B. von, Plachter, H. & Rossler, M. 1995. *Cultural Landscapes of Universal Value*. Jena, Stuttgart, New York: Gustav Fischer Verlag.

Douglas, C.H. 2006. Small Island States and Territories: Sustainable Development Issues and Strategies – Challenges for Changing Islands in a Changing World. *Sustainable Development* 14: 75–80.

Dunlap, R.E. & Catton, W.R. 1992. Towards and Ecological Sociology: The Development, Current Status and Probable Future of Environmental Sociology. *The Annales of the International Institute of Sociology* 3: 262–284.

Durbin, P.T. 1997. Can There be a Best Ethic of Sustainability? *Society for Philosophy and Technology* 3–4(2), http://scholar.lib.vt.edu/eejournals/SPT/v2n2/durbin.html [30.06.2008].

Dyczewski, L. 1995. *Kultura polska w procesie przemian*. Lublin: KUL.

Dylewski, R. 2006. Problemy rozprzestrzeniania sie miast w swietle doswiadczen krajow Unii Europejskiej i Stanow Zjednoczonych. In S. Kozlowski (ed.), *Studia nad zrownowazonym rozwojem*: 27–44. EiS: Bialystok, Lublin, Warsaw.

Earth Summit – Agenda 21. The United Nations Programme of Action from Rio. The Final Text of Agreements negotiated by Governments at the United Nations Conference on Environment and Development (UNCED) 3–14 June 1992. 1992. Rio De Janeiro: UN.

Eckerberg, K. 1997. The Road to Sustainability – The Political History. In S. Sörlin (ed.), *The Road Towards Sustainability, a Historical Perspective*: 36. Uppsala: Uppsala Publishing House.

Eckholm, E.P. 1982. *Down to Earth, Environment and Human Needs*. London: Pluto Press.

Edwards, R. 2002. Japan's Nuclear Safety Dangerously Weak. In *New Scientist Tech* 1 October.

EEAC. 2005. *Draft EEAC Statement on Biodiversity Conservation and Adaptation to the Impact of Climate Change.* http://www.eeac-net.org/download/EEAC%20 Biodiversity%20 statement _27Sep05.pdf [30.06.2007].

Elgin, D. 2006. Voluntary Simplicity and the New Global Challenge. In Haenn N., Wilk R.R. (eds), *The Environment in Anthropology. A Reader in Ecology. Culture and Sustainable Living*: 458–468. New York, London: New York University Press.

Ekins, P., Hutchinson, R.J. and Hillman, M. 1992. *The Gaia Atlas of Green Economics.* New York: Anchor Books.

Encyclopaedia Britannica 2000, CD-ROM.

Endangered Species Act of 1973, as Amended Through the 108th Congress. 2004. Washington: Department of the Interior U.S. Fish and Wildlife Service.

Environmental Law Program. 2004 Year in Review. 2005. San Francisco: Sierra Club.

The Equator Principles. A Financial Industry Benchmark for Determining, Assessing, and Managing Social & Environmental Risk in Project Financing. 2006. http://www. equator-principles.com [30.06.2009].

Esty, D.C. & Gentry, B.S. 1997. Foreign Investment, Globalization and Environment. In T. Jones (ed.), *Globalization and Environment*: 141–172. Paris: OECD.

Europe's Environment: The Second Assessment, an Overview. 1998. Luxembourg: European Environment Agency.

Ferrer-i-Carbonell, A. & Gowdy, J.M. 2007. Environmental Degradation and Hapiness. *Environmental Economics* 60: 509–516.

Ferero J. 2005. The people of this high Andean city were ecstatic when they won the "water war". *New York Times*. December 14.

Ferry, L. 1992. *La nouvel orde écologique. L'arbre, l'animal et l'homme*. Paris: Bernard Grasset.

Fines, F. Ochrona przyrody, fauny i flory. In J.C. Masclet (ed.), *Wspolnota europejska a srodowisko naturalne. Konferencja w Angers*: 254. Lublin: KUL.

Forman, R.T.T. 1984. Corridors in a Landscape: Their Ecological Structure and Function. *Ekologia* 2: 375–385.

Forrester, J.W. 1971. *World Dynamics*. Cambridge: Wright-Allen Press.

Fothergill, A. (producer). 2006. Planet Earth – The Future. London: BBC Films.

Fox-Penner, P. 1997. Environmental Quality, Energy Efficiency and Renewable Energy. In *Electric Utility Restructuring, A Guide to the Competive Era*: 333–369. Vienna: Public Utility Report.

Franczykowska, A. 2006. Znaczenie podmiotow gospodarczych w ksztaltowaniu stosunkow przestrzennych Lubelszczyzny. In S. Kozlowski (ed.), *Zywiolowe rozprzestrzenianie sie miast*: 386–340. Bialystok, Lublin, Warsaw: EiS.

Frankel, G. 2004. Polish Farmers Raise a Stick Over U.S. Agribusiness Giant. *Washington Post* 2 February.

French D. 2002. The Role of the State and International Organization in Reconciling Sustainable Development and Globalization. In *International Environmental Agreements. Politics, Law, and Economics* 2: 135–140.

Friedman, T.L. 2000. *The Lexus and the Olive Tree*. New York: Farrar, Straus and Giroux.

Frissen, P.H.A. 2000. Knowledge and the Betuwe Track. In R.J. Veld (ed.), *Willingly and Knowingly. The Role of Knowledge About Nature and the Environment in Policy Process*: 59–68. Utrecht: Lemma.

Fromm, E. 1976. *Haben oder Sein. Die seelischen Grundlagen einer neuen Gesellschaft.* Stuttgart: Deutsche Verlags-Anstalt.

From Vicious To Vicious Circles? Gender and Micro-Enterprise Development. 1995. Geneva: United Nations Research Institute for Social Development.

Frosch, R.A. 1989. Industrial Ecology: A Philosophical Introduction. In *Proceedings of the National Academy of Sciences*: 800–803. Washington: NAoS.

Futrell, J.W. 2004. Defining Sustainable Development Law. *American Bar Association Journal of Natural Resources and Environment* Autumn:10–11.

Gabbard, A. 1993. Coal Combustion: Nucelar Resource or Danger. *Oakridge National Laboratory Review* 23, Summer/Autumn:

Gallagher, P.K. 2001. Reforming Global Institutions. In J.M. Harris, T.A. Wise, K.P. Gallagher & N.R. Goodwin (eds), *A Survey of Sustainable Development. Social and Economic Dimensions. Frontier Issues in Economic Thought*: 343. Washington, Covelo, London: Island Press.

Galkowski, J.W. 1992. Czlowiek – przyroda – wartosci. In L. Pawłowski & S. Zieba (eds), *Jakiej filozofii potrzebuje ekologia*: 44–45. Lublin: Politechnika Lubelska.

Garcia-Serna, J. Perez-Barrigon, L. & Cocero, M.J. 2007. New Trends for Design Towards Susiainability. *Chemical Engineering Journal* 133: 7–30.

Gardner, G. 2006. *Inspiring Relligons' Contributions to Sustainable Development.* New York: Worldwatch Institute and W.W. Norton & Company.

Garner, A. & Keoleian, G.A. 1995. *Industrial Ecology: An Introduction.* Michigan: National Pollution Prevention Center for Higher Education.

Gawor, L. 2006. Wizja nowej wspolnoty ludzkiej w idei zrownowazonego rozwoju. *Problemy Ekorozwoju/Problems of Sustainable Development* 2(1): 59–66.

Gawor, L. 2008. Globalization and its Alternatives: Antiglobalism, Alterglobalism, and the Idea of Sustainable Development. *Sustainable Development* 2(16): 126–134.

German Advisory Council on Global Change. 2001. *World in Transition. Conservation and Sustainable Use of the Biosphere.* London: Earthscan.

Geise, M. 2000. Ekorozwoj w swietle teorii ekonomii. In A. Papuzinski (ed.), *Polityka ekologiczna III Rzeczypospolitej*: 387–401. Bydgoszcz: Akademia Bydgoska.

Giordano, K. 2006. Metody ekonomicznej wyceny srodowiska jako narzedzie wdrazania rozwoju zrownowazonego. In S. Kozlowski & A. Haladyj (eds), *Rozwoj zrownowazony na szczeblu krajowym, regionalnym i lokalnym*: 64–71. Lublin: KUL, LSB.

Glasby, G.P. 1995. Concept of Sustainable Development: a Meaningful Goal? *The Science of the Total Environment* 1(159): 71–76.

Global Trends 2015: A Dialogue About the Future with Nongovernment Experts. 2000. National Intelligence Council. http://infowar.net/cia/publications/globaltrends2015/ [30.02.2008].

Goncz, E., Skirke, U., Kleizen, H. & Barber, M. 2007. Increasing the Rate of Sustainable Change: a Call for a Redefiniton of the Concept and the Model for its Implementation. *Journal of Cleaner Production* 6(15): 525–537.

Goodland, R. & Daly, H. 1995. Environmental Sustainability. In F. Vanclay & D.A. Bronstein (eds), *Environmental and Social Impact Assessment*: 303–307. Chichester, New York, Brisbane, Toronto, Singapour: Wiley & Sons.

Goodwin, N.R. 2001. Taming the Corporation. In J.M. Harris, T.A. Wise, K.P. Gallagher & N.R. Goodwin (eds), *A Survey of Sustainable Development. Social and Economic Dimensions. Frontier Issues in Economic Thought*: 269. Washington, Covelo, London: Island Press.

Gore, A. 2000. *Earth in the Balance: Ecology and the Human Spirit.* Boston: Houghton Mifflin Co.

Gore, A. 2006. *An Inconvenient Truth: The Planetary Emergency of Global Warming and What We Can Do About It.* New York: Rodale Books.

Gore, Ch. 1995. Markets, Citizenship, and Social Exclusion. In *Social Exclusion: Rhetoris, Reality, Responses*: 1–140. Geneva: International Labor Organization.

Graaf, H.J., Musters, C.J.M. & Keurs, W.J. 1996. Sustainable Development. Looking for New Strategies. *Ecological Economics* 16: 205–216.

Graedel, T.E. & Klee, R.J. 2002. Getting Serious about Sustainability. *Environmental Science & Technology* 4(36): 523–529.

Grzesica, J. 1983. *Ochrona srodowiska naturalnego czlowieka. Problem teologiczno-moralny.* Katowice. Ksiegarnia Sw. Jacka.

Greiff, R. 2005. Zrownowazony rozwoj, zachowanie architektury i ochrona zabytkow. In G. Banse & A. Kiepas (eds), *Zrownowazony rozwoj: od naukowego badania do politycznej strategii*: 229–236. Berlin: Edition Sigma, Berlin.

Greig, A., Hulme, D. & Turner, M. 2007. *Challenging Global Inequality. Development Theory and Practice in the 21st Century*. Basingstoke Hants: Palgrave Macimillan.

Greig, S., Pike, G. & Selby, D. 1987. *Earthrights. Education as if the Planet Really Mattered*. London: WWF, Kogan Page.

Grossarth, S.K. & Hecht, A.D. 2007. Sustainability at the U.S. Environmental Protection Agency. *Ecological Engineering* 30: 1–8.

Grove, R.G. 1992. Rodowod zachodniej polityki ochrony srodowiska. *Swiat Nauki* 9: 18–24.

Grzegorczyk, A. 1992. Kultura i samodyscyplina jako urzeczywistnienie czlowieczenstwa. In L. Pawłowski & S. Zieba (eds), *Jakiej filozofii potrzebuje ekologia*: 19–25. Lublin: Politechnika Lubelska.

Grzesica, J. 2005. Rozwoj a ochrona srodowiska naturalnego czlowieka. In F. Piontek & J. Czerny (eds), *Humanistyczne, ekonomiczne i ekologiczne aspekty kategorii „rozwoj"*: 107. Warsaw, Bytom: KCiS PAN, WSEiA.

Gupta, J. 2002. Global Sustainable Development Governance: Institutional Challenges From a Theoretical Perspective. *International Environmental Agreements Politics. Law and Economics* 2: 367–368.

Hadynska, A. & Hadynski, J. 2006. Conceptions of Multifunctionality: The state-of-the-art in Poland. *European Series on Multifunctionality* 10: 83–107.

Hajduk, Z. 1996. Postep naukowy, techniczny oraz cywilizacyjno-kulturowy. In *Czlowiek i Przyroda* 5: 39–59.

Haliniak, M. 2004. Zrownowazony rozwoj a spoleczenstwo informatyczne. In A. Pawłowski (ed.), *Filozoficzne, spoleczne i ekonomiczne uwarunkowania zrownowazonego rozwoju*: 169–188. Lublin: KIS.

Halaczek, B. 1991. *U progow ludzkosci*. Warsaw: ATK.

Hamm, R. *About IAIA*, In F. Vanclay & D.A. Bronstein (eds). *Environmental and Social Impact Assessment*: xiv. Chichester, New York, Brisbane, Toronto, Singapour: Wiley & Sons.

Hamond, M.J., DeCanio, S.J., Duxbury, P., Sanstad, A.H. & Stinson, Ch. H. 2001. An Idea Whose Time Has Come. In J.M. Harris, T.A.Wise, K.P. Gallagher & N.R. Goodwin (eds), *A Survey of Sustainable Development. Social and Economic Dimensions. Frontier Issues in Economic Thought*: 314. Washington, Covelo, London: Island Press.

Hannenberg, P. 1992. Earth Summit'92 in Rio, Heated Battle over Planet's Survival. *Enviro* 13: 2–5.

Haq, M. 1995. The Human Development Paradigm. In *Reflections on Human Development*. New York: Oxford University Press.

Hall, E.T. 1987. *Bezglosny jezyk*. Warsaw: PIW.

Hall, E.T. 2001. *Ukryty wymiar*. Warsaw: Muza S.A.

Harries, J.E. 2000. The Rainbow Planet Studies of the Earth's Climate From Space. In J. Mason (ed.), *Highlights in Environmental Research. Professorial Inaugural Lectures at Imperial College*: 716. London: Imperial College Press.

Harris, J.M., Wise, T.A., Gallagher, K.P. & Goodwin, N.R. (eds). 2001. *A Survey of Sustainable Development. Social and Economic Dimensions*. Washington, Covelo, London: Island Press.

Harris, J.M. 2001. Economics of Sustainability: The Environmental Dimension. In J.M. Harris, T.A.Wise, K.P. Gallagher & N.R. Goodwin (eds), *A Survey of Sustainable Development. Social and Economic Dimensions. Frontier Issues in Economic Thought*: 5–10. Washington, Covelo, London: Island Press.

Harris, J.M. 2001. Population and Urbanization. Overview Essay. In J.M. Harris, T.A.Wise, K.P. Gallagher & N.R. Goodwin (eds), *A Survey of Sustainable*

Development. Social and Economic Dimensions. Frontier Issues in Economic Thought: 122–125. Washington, Covelo, London: Island Press.

Harris, J.M. & Goodwin, N.R. 2001. Volume Introduction. In J.M. Harris, T.A.Wise, K.P. Gallagher & N.R. Goodwin (eds), *A Survey of Sustainable Development. Social and Economic Dimensions. Frontier Issues in Economic Thought*: XXVII. Washington, Covelo, London: Island Press.

Hawrylyshyn, B. 1980. *Road Maps to the Future. Towards More Effective Societies.* Oxford: Pergamon Press.

Hay, R. 2005. Becoming Ecosynchronous, Achieving Sustainable Development via Personal Development. *Sustainable Development* 14: 1–15.

Hayward, T. 1994. *Ecological Thought.* Cambridge: Polity Press.

Helcel, Z. 1856. *Dawne Prawo Polskie. Starodawne prawa polskiego pomniki.* Warsaw: Helcel.

Hellström, T. 2007. Dimensions of Environmentally Sustainable Innovation: the Structure of Eco-Innovation Concepts. *Sustainable Development* 15: 148–159.

Heredia, C.A. 1996. The World Bank and Poverty. In P. Bosshard (ed.), *Lending Credibility: New Mandates and Partnership for the World Bank* 229–242. Washington: WWF.

Hinterberger, F. & Schmidt-Bleek, F. 1999. Dematerialization, MIPS and Factor 10. Physical Sustainability Indicators as a Social Device. *Ecological Economics* 29: 53–56.

Honey, M. 2006. Treading Lightly? Ecotourism's Impact on the Environment. In N. Haenn & R.R. Wilk (eds), *The Environment in Anthropology. A Reader in Ecology. Culture and Sustainable Living*: 449–457. New York. London: New York University Press.

Hluszczyk, H. & Stankiewicz, A. 1996. *Ekologia.* Warsaw: WSiP.

Holden, E. & Linnerud, K. 2007. The Sustainable Development Area: Satisfying Basic Needs and Safeguarding Ecological Sustainability. *Sustainable Development* 15: 174–187.

Holmberg, J. 1992. *Making Development Sustainable: Redefining Institutions, Policy and Economics.* Washington: Island Press.

Hore-Lacy, I. 2006. *Responsible Dominion, A Christian Approach to Sustainable Development.* Vancouver: Regent College Publishing.

Howard, E. 1902. *Garden Cities of Tomorrow.* http://www.library.cornell.edu/Reps/ DOCS/ howard. htm [30.06.2008].

Huesemann, M.H. 2002. The Inherent Biases in Environmental Research and Their Effects on Public Policy. *Futures* 34: 621–633.

Hull, Z. 1993. Dylematy i wymiary ekorozwoju. *Postepy Nauk Rolniczych* 3: 7–8.

Hull, Z. 1994. Filozoficzne przeslanki i zalozenia edukacji ekologicznej. *Biuletyn Naukowy* 1(13): 22.

Hull, Z. 2003. Filozofia zrownowazonego rozwoju. In A. Pawłowski (ed.), *Filozoficzne i spoleczne uwarunkowania zrownowazonego rozwoju*: 15–26. Lublin: KIS.

Hull, Z. 2003. Filozoficzne podstawy ochrony srodowiska. In Abdank-Kozubski A., Czartoszewski J.W. (eds), *Humanistyczny profil ochrony srodowiska*: 97–104. Warsaw: UKSW.

Hull, Z. 2007. Czy idea zrownowazonego rozwoju ukazuje nowa wizje rozwoju cywilizacyjnego? *Problemy Ekorozwoju/Problems of Sustainable Development* 1(2): 49–57.

Hull, Z. 2008. Filozoficzne i spoleczne uwarunkowania zrownowazonego rozwoju. *Problemy Ekorozwoju/Problems of Sustainable Development* 1(3): 27–31.

Hull, Z. 2008. Sustainable Development: Promises, Understanding and Prospects. *Sustainable Development* 2(16): 73–80.

Hukkinen, J. 2001. Eco-efficiency as Abandonment of Nature. *Ecological Economics* 2001: 311–315.

Hultman, B. 1992. *Water and Wastewater Management in the Baltic Region.* Uppsala: The Baltic University Programme, Uppsala.

Ikerd, J.E. 2005. *Sustainable Capitalism a Matter of Common Sense*. Bloomfield: Kumarin Press Inc.

Irked, J. 2008. Sustainable Capitalism: A Matter of Ethics And Morality. *Problemy Ekorozwoju/Problems of Sustainable Development* 1(3): 13–22.

Ingarden, R. 2001. *Ksiazeczka o czlowieku*. Cracow: WL.

International Energy Annual 2005. 2007. Washington: Energy Information Administration.

Ite U.E. 2007. Partnership with the State for Sustainable Development: Shell's Experience in the Niger Delta, Nigeria. *Sustainable Development* 15: 216–228.

Jackson, J.P. 1996. Unending Nightmare. *Time* 6 May: 23–25.

Jaroszynski, P. 1995. Demokracja – politeja czy ochlokracja. *Filozofia* 24: 7–8.

Jaskowski, J. 1988. Luki i niedomowienia. *Nowy Medyk* 16 December: 8–9.

Jezowski, P. 2000. *Ochrona srodowiska i ekorozwoj*. Warsaw: SGH.

Jezowski, P. 2002. Metoda deklarowanych preferencji na tle metod analizy i wyceny wartosci ekologicznych. In J. Szyszko, J. Rylke & P. Jezowski (eds), *Ocena i wycena zasobow przyrodniczych*: 237–252. Warsaw: SGGW.

Jedrzejczyk, D. 2001. *Wprowadzenie do geografii humanistycznej*. Warsaw: UW.

John Paul II. 1979. *Redemptor hominis*. Vatican. http://www.vatican.va/holy_father/ john_ paul_ii/encyclicals [1.03.2011].

John Paul II. 1987. *Sollicitudo rei socialis*. Vatican. http://www.vatican.va/holy_father/ john_ paul_ii/encyclicals [1.03.2011].

John Paul II. 1989. *Message for the World Day of Peace*. Vatican. http://www.vatican.va/ holy_ father/john_paul_ii/encyclicals [1.03.2011].

John Paul II. 1991. *Centessimus annus*. Vatican. http://www.vatican.va/holy_father/ john_ paul_ii/encyclicals [1.03.2011].

John Paul II. 1995. *Evangelium vitae*. Vatican. http://www.vatican.va/holy_father/john_ paul_ ii/encyclicals [1.03.2011].

John Paul II. 2001. *John Paul II's address to participants in the plenary assembly of the Pontifical Academy of Social Sciences*. Vatican. *http://famvin.org/en/archive/globalization-neither-good-nor-bad-will-be-what-people-make-of-it-pope* [11.03.2009].

Johnson, D.J. 2002. *Sustainable Development: Our Common Future*. http://ww.oecdobserver. org/ news/printpage.php [1.01.2011].

Jonas, H. 1974. *Das Prinzip Verantwortung, Versuch einer Ethik für Technologische Zivilization*. Franfurt: Insel Verlag.

Jonas, H. 1974. *Philosophical Essays. From Ancient Creed to Technological Man*. New Jersey: Englewood Cliffs, New Jersey' Prentice-Hall.

Jonas, H. 1984. *The Imperative of Responsibility. In Search of an Ethics for the Technological Age*. Chicago, London: The University of Chicago Press.

Jones, R., Summers, M. & Mayo E. 1996. *Sustainable Agriculture, Economic Alternatives for Eastern Europe III*. London: The New Economics Foundation.

Juchnowicz, S. 2006. Polski Klub Ekologiczny w 25-lecie dzialalnosci. In S. Kozlowski (ed.), *Zywiolowe rozprzestrzenianie sie miast*: 11–24. Bialystok, Lublin, Warsaw: EiS.

Kalinowska, A. 1992. *Ekologia – wybor przyszlosci*. Warsaw: Editions Spotkania.

Kalinowska, A. 2003. Rola komunikacji spolecznej w zapobieganiu konfliktom wokol ochrony przyrody. In J.W. Czartoszewski (ed.), *Konflikty spoleczno-ekologiczne*. Warsaw: Verbinum.

Kaminski, S. 1992. *Nauka i metoda, pojecie nauki i klasyfikacja nauk*: 109–110. Lublin: KUL.

Karlsson, S. 1997. *Man and Material Flows. Towards Sustainable Material Management*. Uppsala: Uppsala Publishing House.

Kassenberg, A. 2002. *Raport 2/2002 Kompas Rio +10. Spoleczna ocena realizacji przez Polske dokumentow przyjetych na konferencji ONZ „Srodowisko i rozwoj" w czerwcu 1992 r. w Rio de Janeiro*. Warsaw: InRE.

Keating, M. 1993. *Agenda for Change: A Plain Language Version of Agenda 21 and Other Rio Agreements*. Geneva: Centre for Our Common Future.

Kele, F. & Mariot, P. 1986. *Krajobraz, czlowiek, srodowisko*. Wroclaw, Warsaw, Cracow, Gdansk, Lodz: Ossolineum.

Kereiva, P. & Marvier, M. 2007. Przyroda czy czlowiek? *Swiat Nauki* 11: 47–48.

Kiepas, A. 1993. W strone etyki odpowiedzialnosci. *Transformacje* 3: 17.

Kiepas, A. 2006. Etyka jako czynnik ekorozwoju w nauce i technice. *Problemy Ekorozwoju/Problems of Sustainable Development* 2(1): 67–72.

Kihlstrom, J.H. 1992. *Toxicology. The Environmental Impact of Pollutants*. Uppsala: Uppsala Publishing House.

Kiss, A. & Shelton, D. 1993. *Manual of European Environmental Law*. Cambridge: Grotius Publications Inc.

Klein, N. 2000. *No Logo: Taking Aim at the Brand Bullies*. New York: Picador.

Klein, N. 2007. *The Shock Doctrine. The Rise of Disaster Capitalism*. New York: Knopf.

Klemmensen, B., Pedersen, S., Dircknick-Holmfeld, K.R., Marklund, A.A & Ryden., L. 2007. *Environmental Policy. Legal and Economic Instruments*. Upssala: The Baltic University Press.

Knamiller, G.W. 1987. Environmental Education in Schools. In A.V. Baez, G.W. Knamiller & J.C. Smyth (eds), *The Environment, Science and Technology Education and Future Human Needs* 8: 55–57. Oxford, New York, Beijing, Frankfurt, Sao Paulo, Sydney, Tokyo, Toronto: Pergamon Press.

Knercer, W. 2003. Wsie Warmii i Mazur w rozwoju historycznym. In Lizewska I., Knercer W. (eds), *Zachowane-ocalone, o krajobrazie kulturowym i sposobach jego ksztaltowania*: 11–12. Olsztyn: BKB.

Kolakowski, L. 1990. *Cywilizacja na lawie oskarzonych*. Warsaw: Res Public.

Kosmicki, E. 1993. Globalne perspektywy rozwoju gospodarczego i ochrony srodowiska. *Aura* 9: 25.

Kostecka, J., Paczka, G. & Mroczek, J.R. 2007. Zielen miejska jako element zrownowazonego rozwoju miasta. In S. Kozlowski & P. Legutko-Kobus (eds), *Planowanie przestrzenne – szanse i zagrozenia spoleczno-srodowiskowe*: 280. Lublin: KUL.

Kostka, M.S. 1993. Ochrona lasow. In B. Prandecka (ed.), *Interdyscyplinarne podstawy ochrony srodowiska przyrodniczego*: 182–198. Wroclaw, Warsaw, Cracow: Ossolineum.

Kozak, Z. 1993. Marzenia, nadzieje i samospelniajace sie przepowiednie. In S. Kyc (ed.), *Kryzys idei postepu – wymiar ekologiczny*: 167–171. Lublin: Politechnika Lubelska.

Kozak, Z. 2001. Ochrona srodowiska w czasach postmodernizmu. In J.W. Czartoszewski (ed.), *Edukacja ekologiczna na progu XXI wieku*: 202–205. Warsaw: Verbinum.

Kozicki, W. 1913. *W obronie kosciolow i cerkwi drewnianych*. Lviv: C.K. KGW.

Kozlowski, S. 1991. *Gospodarka a srodowisko przyrodnicze*. Warsaw: PWN.

Kozlowski, S. 1993. Zalozenia ery ekologicznej. *Problemy Ekorozwoju/Problems of Sustainable Development* 4–5: 2–9.

Kozlowski, S. 2005. *Przyszlosc ekorozwoju*. Lublin: KUL.

Kozo, M. 1994. Development, Ecological Degradation and North-South Trade. In M. Jankowska-Kramer, M. Urbaniec & A. Krynski (eds), *Miedzynarodowe zarzadzanie srodowiskiem*. Warsaw: C.H. Beck.

Krakowiak, J.L. 1997. *Ziemia domem czlowieka*. Warsaw: PTU.

Kraoll, G. 2006. *Rachel Carson's Silent Spring. A Brief History of Ecology as a Subversive Subject*. http://www.onlineethics.org/cms/9174.aspx [28.02.2011].

Krapiec, M.A. 1990. Prawda – dobro – piekno jako wartosci humanistyczne. In B. Suchodolski (ed.), *Alternatywna pedagogika humanistyczna*: 45. Wroclaw, Warsaw, Cracow: Ossolineum.

Krapiec, M.A., Kaminski, S., Zdybicka, Z.J. & Jaroszynski, P. 1992. *Wprowadzenie do filozofii*. Lublin: KUL.

Kras, E. 2007. *The Blockage. Rethinking Organizational Principles for the 21st Century*. Baltimore: American Literary Press.

Krebs, Ch.J. 1998. *Ecology*. Menlo Park: Benjamin Cummings.

Krikke, B. & Zaworska-Matuga, W. 2001. *Planowanie i wdrazanie polityki ochrony srodowiska*. Lublin: El-Press.

Krishna-Hensel, S.F. 1999. Population and Urbanization in the Twenty-First Century: India's Megacities. In B. Baudot & W.R. Moomaw (eds), *People and Their Planet: Searching for Balance*: 157–173. London: Macmillan.

Krzysztofek, K. 2004. *Kultura polska wobec integracji europejskiej i globalnego spoleczenstwa informacyjnego*. Warsaw: FRP, IWoS.

Kucowski, J., Laudyn, D. & Przekwas, M. 1993. *Energetyka a ochrona srodowiska*. Warsaw: WNT.

Kuderowicz, Z. 1993. Odpowiedzialnosc za srodowisko. In B. Prandecka (ed.), *Interdyscyplinarne podstawy ochrony srodowiska przyrodniczego*: 17–23. Wroclaw, Warsaw, Cracow: Ossolineum.

Kukulka, J. 1992. Internacjonalizacja problemow ekologicznych a ksztaltowanie ladu miedzynarodowego. In E.J. Palyga (ed.), *Ekologia spoleczna i wspolpraca miedzynarodowa w zakresie ochrony srodowiska*: 120–121. Warsaw: AKEE.

Kyoto Protocol to the United Nations Framework Convention on Climate Change. 1998. New York: United Nations.

Lafferty, W.M. & Meadowcroft, J. 2000. *Implementing Sustainable Development. Strategies and Initiatives in High Consumption Societies*. New York: Oxford University Press: New York.

Lang, S. 1998. U.S. Floating in Stinky Problem: Manure Pollution. *Desert New* 29 April.

Laszlo, C. 2005. *The Sustainable Company. How to Create Lasting Value Though Social and Environmental Performance*. Washington, Covelo, London: Island Press.

Laszlo, E. 1977. *Goal for Mankind, a Report to the Club of Rome on the New Horizons of Global Community*. New York: Dutton.

Lawrence, R.F. & Thomas, W.L. 2004. The Equator Principles and Project Finance: Sustainability in Practise? *American Bar Association Journal of Natural Resources and Environment* Autumn: 20–26.

Lean, G. & Hinrichsen, D. 1992. *Atlas of the Environment*. New York: Harper Perennial.

Legocki, A.B. 2006. Ziemia na rozdrozu – wyzwania i dylematy ery biologii. *Nauka* 4: 39.

Leistritz, F.L. 1995. Economic and Fiscal Impact Assessment. In F. Vanclay & D.A. Bronstein (eds), *Environmental and Social Impact Assessment*: 129–139. Chichester, New York, Brisbane, Toronto, Singapour: Wiley & Sons.

Lenart, W. 2008. *Zakres mozliwej wspolpracy Bialorusi, Litwy, Polski, Rosji i Ukrainy i w dziedzinie gospodarki przestrzennej, zrownowazonego rozwoju oraz promocji regionalnej*. http:// www.fzpp.pl/assets/files/Referat%20WL.doc [30.01.2009].

Lenkowa, A. 1981. *Aby swiat nie stal sie pustynia, karty z historii ochrony przyrody*. Warsaw: KAW.

Leopold, A. 1993. The Land Ethic. In Armstrong S.J. & Botzler R.G. (eds), *Environmental Ethics. Divergence and Convergence*: 373–382. New York: McGraw-Hill Inc.

Leroy, P. & Nelissen, N. 1999. *Social and Political Sciences of the Environment, Three Decades of Research in the Netherlands*. Utreht: International Books.

Leva, Ch. E. Di. 2004. Sustainable Development and the World Bank's Millennium Development Goals. *American Bar Association Journal of Natural Resources and Environment* Autumn 2004: 14.

Lin, G.T.R. 2007. Empirical Measurement of Sustainable Welfare from the Perspective of Extended Genuine Savings. *Sustainable Development* 15: 188–203.

Lipietz, A. 1995. Enclosing the Global Commons: Global Environmental Negotiations in a North-South Conflictual Approach. In V. Bhaskar & A. Glyn (eds), *The North, The South, and the Environment*: 118–145. New York: St. Martin's Press.

Lipinska, B. 2003. Kultura uzytkowania przestrzeni – degradacja krajobrazu wiejskiego. In I. Lizewska & W. Knercer (eds), *Zachowane-ocalone, o krajobrazie kulturowym i sposobach jego ksztaltowana*: 132. Olsztyn: SBKB.

Lipton, M. 1997. Accelerated Resource Degradation by Agriculture in Developing Countries, The Role of Population Change and Responses to it. In S.A. Vosti & T. Reardon (eds), *Agroecological Perspective* 79–89. Baltimore: John Hopkins University Press.

Liro, A.1995. *Koncepcja krajowej sieci ekologicznej ECONET – POLSKA*. Warsaw: IUCN Poland.

The Lisbon Strategy. 2000. EC: Brussels http://europa.eu/scadplus/glossary/lisbon_strategy/en.htm [30.06.2009].

Lisicka, H. 2000. *Polityka w ochronie srodowiska w polityce panstwa*. In A. Papuzinski (ed.), *Polityka ekologiczna III Rzeczypospolitej*: 278. Bydgoszcz: Akademia Bydgoska.

Lisiecka, H., Macek, I. & Radecki, W. 1999. *Leksykon ochrony srodowiska – prawo i polityka*. Wroclaw: TNPOS.

Liszewski, D. 2007. Etyczne podstawy rozwoju zrownowazonego. *Problemy Ekorozwoju/ Problems of Sustainable Development* 1(2): 27–33.

Littig, B. & Griesler, E. 2005. Social Sustainability: A Catchword Between Political Pragmatism and Social Theory. *International Journal of Sustainable Development* 1–2(8): 65–79.

Liu, F. 2001. *Environmental Justice Analysis, Theories, Methods and Practice*. Boca Raton, London, New York: Lewis Publishers.

Livermann, D.M., Hanson, M.E., Brown, B.J. & Merideth, Jr. R.W. 1988. Global Sustainability: Toward Measurement. *Environmental Management* 12: 133–143.

Living Planet Report 2006. 2006. Gland: WWF.

Lizewska, I. & Knercer, W. 2003, *Zachowane-ocalone, o krajobrazie kulturowym i sposobach jego ksztaltowania*. Olsztyn: SBKB.

Lothigius, J. 1995. Conserving Biodiversity Helps in Fight Against Poverty. *Enviro* 19: 17–18.

Lovelock, J. 1989. *Gaia: A New Look at Life on Earth*. Oxford: Oxford University Press.

Lovelock, J. 1995. *The Ages of Gaia: a Biography of our Living Earth*. New York, London: W.W. Norton.

Lovins, A.B. 1977. *Soft Energy Paths. Toward A Durable Peace*. London: Friends of Earth International, Penguin Books.

Lovins, L.H. 2004. Natural Capitalism: Path to Sustainability? *American Bar Association Journal of Natural Resources and Environment* Autumn: 6.

Luken, R.A. 2006. Where is Developing Country Industry in Sustainable Development Planning? *Sustainable Development* 14: 46–61.

Luken, R.A. & Hesp, P. 2007. The Contribution of Six Developing Countries's Industry to Sustainable Development. *Sustainable Development* 15: 243–253.

Luttwak, E. 1999. *Turbo-Capitalism: Winners and Losers in the Global Economy*. New York: Harper Collins.

Lapinski, J.L. 2007. Krajobraz kulturowy a ksztaltowanie przestrzenne. In S. Kozlowski & P. Legutko-Kobus (eds), *Planowanie przestrzenne, szanse i zagrozenia spoleczno-srodowiskowe*: 199. Lublin: KUL.

Lozinski, K. 1995. Zaproszenie dla Greenpeace. *Gazeta Wyborcza* 18 October: 9.

Mackenzie, F.T. & Mackenzie, J.A. 1995. *Our Changing Planet. An Introduction to Earth System Science and Global Environmental Change*. New Jersey: Prentice Hall Inc.

Mackow, J., Paczosa, A. & Skirmuntt, G. 2004. *Eko-generacja przyszlosci*. Katowice, Warsaw: WNS.

Mann, H. & Moltke, K. von. 1999. *NAFTA's Chapter 11*. Ottawa: International Institute for Sustainable Development.

Margul, T. 1986. *Religia a przestrzen i krajobraz*. Cracow: UJ.

Marshall, J.D. & Toffel, M.W. 2005. Framing the Elusive Concept of Sustainability: A Sustainability Hierarchy. *Environmental Science & Technology* 3(39)9.

Marten, G.G. 2001. *Human Ecology. Basic Concepts for Sustainable Development*. London: Earthscan.

Martinez-Alier, J. 2001. From Political Economy to Political Ecology. In J.M. Harris, A.T. Wise, K.P.Gallagher & N.R. Goodwin (eds), *A Survey of Sustainable Development. Social and Economic Dimensions. Frontier Issues in Economic Thought*: 32. Washington, Covelo: Island Press.

Marzec, A. 2008. Swiatowa konferencja klimatyczna na Bali. *Czysta Energia* 1: 23.

Mason, J. 2000. *Highlights in Environmental Research. Professorial Inaugural Lectures at Imperial College*. London: Imperial College Press.

Matczak, P. 2000. *Problemy ekologiczne jako problemy spoleczne*. Poznan: WN UAM.

May, R.M. 1992. Ile gatunkow zamieszkuje Ziemie? *Swiat Nauki* 12(16): 20–27.

Mayor, F. & Binde, J. 2001. *The World Ahead: Our Future in the Making*. London: Zed Books & UNESCO Publishing.

Mayoux, L. 1995. What Is Micro-Enterprise Development for Women? Widening the Agenda. In *From Vicious To Vicious Circles? Gender and Micro-Enterprise Development*: 50–58. Geneva: United Nations Researech Institute for Social Development.

Malinowski, A. & Strzalko, J. 1985. *Antropologia*. Warsaw, Poznan: PWN.

Matlack, C. 2011. *A Consortium Wants to Invest $ 560 Billion in Sahara Solar Panels*. http://www. businessweek.com/magazine/content/10_38/b4195012469892.htm [1.03.2011].

Mazurkiewicz, J. 1995. Skonczonosc przyrody i dominacja postaw konsumpcyjnych. *Aura* 6: 4–5.

Meadows, D.H., Meadows, D.L. & Behrens, W.W. 1972. *The Limits to Growth*. New York: Universe Books.

Meadows, D.H., Meadows, D.L. & Randers, J. 1992. *Beyond the Limits: Confronting Global Collapse, Envisioning a Sustainable Future*. Vermount: Chelsea Green Publishing Company, White River Junction.

Meadows, D.H., Randers, J. & Meadows, D.L. 2004. *Limits to Growth, The 30-Years Update*. Vermount: Chelsea Green Publishing Company.

Meara Sheehan, M.O. 2001. Reinventing Cities for People and the Planet. In J.M. Harris, T.A. Wise, K.P. Gallagher & N.R. Goodwin (eds), *A Survey of Sustainable Development. Social and Economic Dimensions. Frontier Issues in Economic Thought*: 151. Washington, Covelo, London: Island Press.

Meinhold, B. 2011. *World's Largest Solar Project Planned for Saharan Desert*. http://inhabitat.com/worlds-largest-solar-project-sahara-desert/ [1.03.2011].

Mersarovic, M. & Pestel, E. 1975. *Mankind at the Turning Point*. New York: Dutton.

Meuleman, L., Niestroy, I. & Hey, C. 2003. *Environmental Governance in Europe*. Utreht: Lemma.

Michael, A.J., Butler, C.D. & Folke, C. 2003. New Visions for Addressing Sustainability. *Science* 302: 1919–1920.

Michna, W. *Ochrona gleb*. 1993. In B. Prandecka (ed.), *Interdyscyplinarne podstawy ochrony srodowiska przyrodniczego*: 165–181. Wroclaw, Warsaw, Cracow: Ossolineum.

Micklethwait, J. & Wooldridge, A. 2000. *A Future Perfect: The Challenge and Hidden Promise of Globalization*. New York: Crown Business.

Michajlow, W. 1978. *Ochrona i ksztaltowanie srodowiska przyrodniczego*. Warsaw: PWN.

Michalowski, A. 2002. Krajobraz kulturowy jako dobro spoleczne. In A. Michalowski A. (ed.), *Krajobraz kulturowy Lodzi i wojewodztwa lodzkiego w nauczaniu mlodziezy*: 7. Warsaw: OOZK.

Michnowski, L. 1995. *Jak zyc?, Ekorozwoj albo …*. Bialystok: EiS.

Michnowski, L. 2003. O potrzebie budowy informacyjnych podstaw trwalego rozwoju w polskiej, europejskiej i swiatowej spolecznosci: 107–120. In A. Pawłowski (ed.), *Filozoficzne i spoleczne uwarunkowania zrownowazonego rozwoju*. Lublin: KIS.

213

Michnowski, L. 2004. Globalizacja inkluzywna jako warunek trwalego rozwoju. In A. Pawłowski (ed.), *Filozoficzne, spoleczne i ekonomiczne uwarunkowania zrownowa-zonego rozwoju*: 189–218. Lublin: KiS.

Michnowski, L. 2007. *Spoleczenstwo przyszlosci a trwaly rozwoj*. Warsaw: KPP2000P PAN.

Michnowski, L. 2011. *Backcasting w projektowaniu strategii sustainable development*, http://www.pte.pl/pliki/2/12/backcasting.ppt [1.02.2011].

Mihelcic, J.R., Philips, L.D. & Watkins, D.W. 2006. Integrating a Global Perspective into Education and Research: Engineering Internatonal Sustainable Development. *Environmental Engineering Science* 3(23): 428.

MicMichael, A.J., Butler, C.D. & Folke, C. 2003. New Visions for Addressing Sustain-ability. *Science* 302: 1919–1920.

Minisch, J. 1995. Ekologiczne sterowanie w sluzbie zrownowazonego rozwoju. In P. Buczkowski (ed.), *Polityka lokalna w zakresie ochrony srodowiska*. Poznan: UAM.

Mizielinski, B. 2002. Wybrane problemy inzynierii srodowiska wewnetrznego. In L. Pawłowski (ed.), *Inzynieria srodowiska, stan obecny i perspektywy rozwoju*: 233–234. Lublin: KIS.

Moberg, A. 1986. *Before and After Chernobyl. Nuclear Power in Crisis. A Country by Country Report*. Malmo: Team Offset.

Moffat, I. 1996. *Sustainable Development: Principles, Analysis and Policies*. New York, London: The Phatrenon Publishing Group.

Moffatt, I. 2000. Ecological Footprints and Sustainable Development. *Ecological Eco-nomics* 32: 359–362.

Monhemius, A.J. 2000. Man, Materials and Environment. In J. Mason (ed.), *Highlights in Environmental Research, Professorial Inaugural Lectures at Imperial College*: 68. London: Imperial College Press.

Moss, D., Davies, C. & Roy, D. 1995. *CORINE Biotopes Sites, Database Status and Perspectives*. European Topic Centre on Nature Conservation, European Environ-ment Agency. http://reports.eea.europa. eu/92-9167-054-5/en [30.06.2009].

Mossakowska, E. 2005. Zmiany poziomu i struktury zatrudnienia ludnosci na obszarach wiejskich w Polsce. *Stowarzyszenie Ekonomistow Rolnictwa i Agrobiznesu* 4(7): 289.

Mucha-Leszko B. 2005. *Wspolczesna gospodarka swiatowa, glowne centra gospodarcze*. Lublin: UMCS.

Munda, G. 1997. Environmental Economics, Ecological Economics and the Concept of Sustainable Development. *Environmental Values* 6: 213–233.

Myers, N. 1986. *The Sinking Ark. A New Look at the Problem of Disappearing Species*. Oxford, New York, Toronto, Sydney, Paris, Frankfurt: Pergamon Press.

Myczkowski, Z. 2001. Ochrona przyrody i ochrona zabytkow – konflikt czy wspol-dzialanie. In K. Pawlowska (ed.), *Architektura krajobrazu a planowanie przestrzenne*: 136–137. Cracow: PK.

Mysliwski, G. 1999. *Czlowiek sredniowiecza wobec czasu i przestrzeni*. Warsaw: Krupski.

Naess, A. 1993. *Ecology, Community & Lifestyle. Outline of an Ecosophy*. Cambridge: Cambridge University Press.

Nagpal, T. & Foltz, C. 1995. *Choosing Our Future. Visions of a Sustainable World*. Baltimore: World Resource Institute.

The Need For Change. The Legacy of TMI: Report of the President's Commission on the Accident at Three Miles Island. 1979. The Commission on the Accident: Washington.

Netboy, A. 1968. *The Atlantic Salmon: a Vanishing Species?* Boston: Houghton Mifflin Co.

New Details on Japan Nuclear Accident. 1999. *Science Daily* 6 December.

Newman, L. 2007. The Virtuous Cycle: Incremental Changes and a Process-Based Sustainable Development. *Sustainable Development* 15: 274–276.

Ney, R. 1994. O wsparcie nauki dla zrownowazonego rozwoju. *Nauka* 2: 123.

Niestroy, I. 2005. *Sustaining Sutainability, a Benchmark Study on National Strategies Towards Sustainable Development and the Impact of Councils in 9 EU Member States*. Den Haag: EEAC.

Nilsson, L., Persson, P.O., Ryden, L., Darozhka, S. & Zaliauskiene, A. 2007. *Cleaner Production, Technologies and Tools for Resource Efficient Production*. Uppsala: The Baltic University Press.

Norton, B.G. 1993. Environmental Ethics and Weak Anthropocentrism. In S.J. Armstrong & R.G. Botzler (eds), *Environmental Ethics. Divergence and Convergence*: 286–288. New York: McGraw-Hill Inc.

Nowak, W. & Stachel, A.A. 2007. *Ocena mozliwosci korzystania z energii wiatru w Polsce na tle krajow Europy i Swiata*, http://www.fundacjarozwoju.szczecin.pl/biuro/teksty2/FRPZ%20%20Referat%20 Nowak+Stachel.pdf [30.01.2011].

No Water, no Future. 2004. *Environmental Science & Technology* 1: 279A.

O'Neill, P. 1993. *Environmental Chemistry*. New York: Chapman and Hall.

O'Riordan, T. 1981. *Environmentalism*. London: Pion Limited.

O'Riordan, T. & Cameron, J. 1994. *Interpreting the Precautionary Principle*. London: Earthscan.

Orr, B. 1992. *The Global Economy*. New York: New University Press.

Ortolano, L. & Shepherd, A. 1995. Environmental Impact Assessment. In F. Vanclay & D.A. Bronstein (eds), *Environmental and Social Impact Assessment*: 3–30. Chichester, New York, Brisbane, Toronto, Singapour: Wiley & Sons.

Our Future, Our Choice. 2001. EC: Luxembourg.

Paczuski, R. 2002. Zrownowazony rozwoj a stosunek polityki do prawa w swietle polskiej praktyki legislacyjnej. In K. Rowny & J. Jablonski (eds), *Zasada zrownowazonego rozwoju w prawie i praktyce ochrony srodowiska* 33–42. Warsaw: WPWSBA.

Page, T. 1997. On the Problem of Achieving Efficiency and Equity, Intergenerationally. *Land Economics* 4(73): 580–596.

Papuzinski, A. 1998. *Zycie – Nauka – Ekologia. Prelegomena do kulturalistycznej filozofii ekologii*. Bydgoszcz: WSP.

Papuzinski, A. 1999. Ekofilozofia jako filozofia ekologii holistycznej. In J.M. Dolega & J.W. Czartoszewski (ed.), *Ochrona srodowiska w filozofii i teologii*: 28. Warsaw: ATK.

Papuzinski, A. 2004. Filozoficzne aspekty zasady zrownowazonego rozwoju a Iustica Socialis. In A. Pawłowski (ed.), *Filozoficzne, spoleczne i ekonomiczne aspekty zrownowazonego rozwoju*: 56–59. Lublin: KIS.

Papuzinski, A. 2008. The Philosophical Dimension to the Principle of Sustainable Development in the Polish Scientific Literature. *Sustainable Development* 2(16): 109–116.

Paryjczak, T. 2008. Promowanie zrownowazonego rozwoju przez Zielona Chemie. *Problemy Ekorozwoju/Problems of Sustainable Development* 1(3): 39–51.

Pasierbiak, P. & Kuspit, J. 2005. Rozwoj wymiany handlowej miedzy Stanami Zjednoczonymi i Unia Europejska w warunkach wspolpracy transatlantyckiej. In B. Mucha-Leszko (ed.), *Wspolczesna gospodarka swiatowa, glowne centra gospodarcze*: 229. Lublin: UMCS.

Paul VI. 1967. *Populorum progresio*. Vatican, http://www.vatican.va/holy_father/ paul_vi/ encyclicals [1.03.2011].

Pawlowska, K. 2001. *Idea swojskosci krajobrazu kulturowego*. Sosnowiec: WNoZ US.

Pawlowska, K. & Swaryczewska, M. 2002. *Ochrona dziedzictwa kulturowego, Zarzadzanie i partycypacja spoleczna*. Cracow: UJ.

Pawlowska, M. 1999. *Mozliwosc zmniejszenia emisji metanu z wysypisk na drodze jego biochemicznego utleniania w rekultywacyjnym nadkladzie glebowym – badania modelowe*. Lublin: Politechnika Lubelska.

Pawłowski, A. 1996. Perception of Environmental Problems by Young People in Poland. *Environmental Education Research* 3(2): 279–285.

Pawłowski, A. 1999. *Odpowiedzialnosc czlowieka za przyrode*. Lublin: Politechnika Lubelska.

Pawłowski, A. 2000. Ekologia – kultura – rodzina. In J.M. Dolega & J.W. Czartoszewski (eds), *Rodzina ludzka w nauce i kulturze*: 311–316. Olecko: WM.

Pawłowski, A. 2001. Dylematy rozwoju zrownowazonego. In A. Pawłowski & M.R. Dudzinska M.R. (eds), *Zrownowazony rozwoj w polityce i badaniach naukowych*: 135–146. Lublin: KCiS PAN.

Pawłowski, A. 2002. Ochrona zabytkow kultury i sztuki, a ochrona „zabytkow" przyrody – wybrane zagadnienia. In J.M. Dolega (ed.), *Podstawy kultury ekologicznej*: 205–211. Warsaw: UKSW.

Pawłowski, A. 2003. Introducing Sustainable Development – a Polish Perspective. In L. Pawłowski., M.R. Dudzinska & Pawłowski, A. (eds), *Evironmental Engineering Studies, Polish Research on the Way to the EU*: 367–375. New York, Boston, Dordrecht, London, Moscow: Kluwer Academic/Plenum Press.

Pawłowski, A. 2004. Wielowymiarowosc rozwoju zrownowazonego. In A. Pawłowski (ed.), *Filozoficzne, spoleczne i ekonomiczne uwarunkowania zrownowazonego rozwoju*: 107–129. Lublin: KIS.

Pawłowski, A. 2005. Edukacja srodowiskowa dla zrownowazonego rozwoju – wybrane problemy. In J.W. Czartoszewski, E. Grzegorzewicz & A.W. Swiderski (eds), *Problemy XXI wieku, prawo ochrony srodowiska, edukacja srodowiskowa i agrobiznes*: 157–165. Warsaw: UKSW.

Pawłowski, A. 2006. The Historical Aspect to the Shapping of the Sustainable Development Concept. In L. Pawłowski, M.R. Dudzinska & A. Pawłowski (eds), *Environmental Engineering*: 21–30. London: Taylor & Francis Group.

Pawłowski, A. 2006. Wielowymiarowosc zrownowazonego rozwoju. *Problemy Ekorozwoju/ Problems of Sustainable Development* 1(1): 23–32.

Pawłowski, A. 2007. Bariery we wprowadzaniu zrownowazonego rozwoju – spojrzenie ekofilozofa. *Problemy Ekorozwoju/Problems of Sustainable Development* 1(2): 59–65.

Pawłowski, L. 2007. Environmental Engineering in the Protection and Management of the Human Environment. In L. Pawłowski, M.R. Dudzinska & A. Pawłowski (eds), *Environmental Engineering*: 3–6. London: Taylor & Francis Group.

Pawłowski, A. 2008. How Many Dimensions Does Sustainable Development Have? *Sustainable Development* 2(16): 81–90.

Pawłowski, A. 2009. Sustainable Energy as a "sine qua non" Condition for the Achievement of Sustainable Development. *Problemy Ekorozwoju/Problems of Sustainable Development* 2(4): 9–12.

Pawłowski, A. 2009. The Sustainable Development Revolution. *Problemy Ekorozwoju/ Problems of Sustainable Development"* 1(4): 65–76.

Pawłowski, A. 2010. Sustainable Development as a Civilisational Revolution. In Z. Lepko & R.F. Sadowski (eds), *A Humanist Approach to Sustainable Development*: 169–184. Warsaw: UKSW.

Pawłowski, A. 2010. Sustainable Development vs Environmental Engineering: Energy Issues. In J. Nathwani & A. Ng, (eds). *Paths to Sustainable Energy*: 13–28. Rijeka: InTech.

Pawłowski, A. 2010. The Role of Environmental Engineering in Introducing Sustainable Development. *Ecological Chemistry and Engineering S* 3(17): 263–278.

Pawłowski, A. & Pawłowski, L. 2004. Realizacja zasady zrownowazonego rozwoju w przemysle cementowym. In *Ochrona i inzynieria srodowiska, zrownowazony rozwoj*: 277–286. Cracow: KIS.

Pawłowski, A. & Pawłowski, L. 2008. Zrownowazony rozwoj we wspolczesnej cywilizacji, czesc 1: srodowisko a zrownowazony rozwoj. *Problemy Ekorozwoju/Problems of Sustainable Development* 1(3): 53–65.

Pawłowski, L., Dudzinska, M.R. & Pawłowski, A. 2006. *Environmental Engineering vol 1*. London: Taylor & Francis Group.

Pawłowski, L., Dudzinska, M.R. & Pawłowski, A. 2007. *Environmental Engineering vol. 2*. London: Taylor & Francis Group.

216

Pawłowski, L., Dudzinska, M.R. & Pawłowski, A. 2010. *Environmental Engineering vol. 3*. Boca Raton, London, New York, Leiden: CRC Press, Taylor & Francis Group, A Balkema Book.

Pawłowski, L. & Pawłowski, A. 2010. Environmental Engineering as a Toll for Managing the Human Environment. In L. Pawłowski, M.R. Dudzinska & A. Pawłowski (eds), *Environmental Engineering vol. 3*: 3–6. Boca Raton, London, New York, Leiden: CRC Press, Taylor & Francis Group, A Balkema Book.

Peski, W. 1999. *Zarzadzanie zrownowazonym rozwojem miast*. Warsaw: Arkady.

Pfahl, S. 2005. Institutional Sustainability. *International Journal of Sustainable Development* 1–2(8): 81–85.

Piatek, Z. 1998. *Etyka srodowiskowa, nowe spojrzenie na miejsce czlowieka w przyrodzie*. Cracow: Ksiegarnia Akademicka.

Piatek, Z. 2006. Bioetyka wobec wyzwan globalizacji. *Problemy Ekologii* 3(10): 123.

Piatek, Z. 2007. *Pawi ogon, czyli o biologicznych uwarunkowaniach kultury*. Cracow: UJ.

Piatek, Z. 2007. Przyrodnicze i spoleczno-historyczne warunki rownowazenia ladu ludzkiego swiata. *Problemy Ekorozwoju/Problems of Sustainable Development* 2(2): 5–18.

Piatek, Z. 2008. Ecophilosophy as a Philosophical Underpinning of Sustainable Development. *Sustainable Development* 2(16): 91–99.

Piatek, Z. 2008. *Ekofilozofia*. Cracow: UJ.

Piontek, B. 2002. *Koncepcja rozwoju zrownowazonego i trwalego Polski*. Warsaw: PWN.

Piontek, F. 2001. Kontrowersje i dylematy wokol rozwoju zrownowazonego i trwalego. In F. Piontek (ed.), *Ekonomia a rozwoj zrownowazony*: 24. Bialystok: EiS.

Piontek, F. 2002. *Kapital ludzki w procesie globalizacji a w zrownowazonym rozwoju*. Wisla: ATH, WSEiA.

Piontek, F. & Piontek, W. 2002. *Rachunek ekonomiczny w ochronie srodowiska*. Bytom: WSEiA.

Piontek, F. 2007. Teoria rozwoju a personologiczna koncepcja teorii ekonomicznej. In B. Piontek & F. Piontek (eds), *Zarzadzanie rozwojem: aspekty spoleczne, ekonomiczne i ekologiczne*: 58. Warsaw: PTE.

Piontek, F. & Piontek, B. 2005. Kategoria kapitalu w warunkach zrownowazonego rozwoju i spoleczenstwa wiedzy. In B. Poskrobko & S. Kozlowski (eds), *Studia nad zrownowazonym rozwojem*: 67–84. Bialystok, Warsaw: KCiS.

Piontek, F. & Piontek, B. 2005. Kategoria „rozwoj" w alternatywnych koncepcjach jego urzeczywistniania. In F. Piontek & J. Czerny (eds), *Humanistyczne, ekonomiczne i ekologiczne aspekty kategorii „rozwoj"*: 30. Warsaw, Bytom: KCiS PAN, WSEiA.

Pinstrup-Andersen, P. & Pandaya-Lorch, R. 1998. Food Security and Sustainable Use of Natural Resources: A 2020 Vison. *Ecological Economics* 26: 1–10.

Platje, J. 2006. Poziom dochodu narodowego a priorytety w rozwoju zrownowazonym. In S. Kozlowski & A. Haladyj (eds), *Rozwoj zrownowazony na szczeblu krajowym, regionalnym, lokalnym – doswiadczenia polskie i mozliwosci ich zastosowania na Ukrainie*: 37–46. Lublin: KUL, LSB.

Poltorak, M. 2006. Zrownowazona produkcja i konsumpcja – zarzadzanie srodowiskiem, analiza cyklu zycia. *Czystsza Produkcja i Eko-zarzadzanie* 4(39): 19–22.

Ponting, C. 1993. *A Green History of the World: The Environment and the Collapse of Great Civilizations*. New York: Penguin Books, New York.

Population Reference Bureau. 2011. *2010 World Population Data Sheet*. Washington: PRB.

Porter, A.L. 1995. Technology Assessment. In F. Vanclay & D.A. Bronstein (eds), *Environmental and Social Impact Assessment*: 67–81. Chichester, New York, Brisbane, Toronto, Singapour: Wiley & Sons.

Poskrobko, B. 1998. *Sterowanie ekorozwojem*. Bialystok: Politechnika Bialostocka.

Poskrobko, B. 1998. Podstawy polityki ekologicznej. In K. Gorka & B. Poskrobko (eds), *Ochrona srodowiska. Problemy spoleczne, ekonomiczne i prawne*: 75. Warsaw: PWE.

217

Poskrobko, B. & Kozlowski, S. *Studia nad zrownowazonym rozwojem*, KCiS, Bialystok, Warsaw 2005.

Poskrobko, B. & Olenska, J. 2001. Regionalne i lokalne strategie rozwoju oraz trwalego i zrownowazonego rozwoju. In F. Piontek, *Ekonomia a rozwoj zrownowazony*: 36–42. Bialystok: EiS.

Postman, N. 1992. *Technopoly the Surrender of Culture to Technology*. New York: Vintage Books, A Division of Random House Inc.

Prandecka, B.K. 1991. *Nauki ekonomiczne a srodowisko przyrodnicze*. Warsaw: PWE.

Radecki, W. 1989. Articles on History of Protection of the Enviornment in Poland. *Aura* 4,5,6,7: 17;16; 22–23.

Radziejowski, J. 2006. Bledy w polityce przestrzennej, a rozprzestrzenianie sie miast w Polsce. In S. Kozlowski (ed.). *Studia nad zrownowazonym rozwojem*: 81–93. Bialystok, Lublin, Warsaw: EiS.

Rassafi, A.A., Poorzahedy, H. & Vaziri, M. 2006. An Alternative Definiton of Sustainable Development Using Stability and Chaos Theories. *Sustainable Development* 14: 65.

Redman, Ch. 2006.The Growth of World Urbanism. In N. Haenn, N. & R.R. Wilk (eds), *The Environment in Anthropology. A Reader in Ecology. Culture and Sustainable Living*. New York, London: New York University Press.

Reed, D. 1997. *Structural Adjustment. The Environment and Sustainable Development*. London: Earthscan.

Rees, W.E. 1999. Consuming the Earth: the Biophysics of Sustainability. *Ecological Economics* 29: 23–27.

Regan, T. 1993. The Case for Animal Rights. In S.J. Armstrong & R.G. Botzler (eds), *Environmental Ethics. Divergence and Convergence*: 321–328. New York: McGraw-Hill Inc.

Reid, W.V., Laird, S.A., Meyer, C.A., Gamez, R., Sittenfeld, A., Janzen, D.H., Gollin, M.A. & Juma, C. 1993. *Biodiversity Prospecting. Using Genetic Resources for Sustainable Development*. Baltimore: World Resource Institute.

Renewed EU Strategy for Sustainable Development. 2006. UE: Brussels. http://ec.europa. eu/ sustainable/ [30.06. 2007].

Repetto, R.C. 1991. Reform of the National Accounting System. In Z. Bochniarz & R. Bolan (eds), *Designing Institutions for Sustainable Development: a New Challenge for Poland*: 185–200. Minneapolis, Bialystok: BTU.

Reut, A. 2000. Powstanie i znaczenie ekonomii ekologicznej. In A. Papuzinski (ed.), *Polityka – ekologia – kultura. Spoleczne przeslanki i przejawy kryzysu ekologicznego*: 109–116. BydGoszcz: WSP.

Richling, A. 2004. Z metodyki klasyfikacji krajobrazu. In M. Kucharczyk (ed.), *Wspolczesne problemy ochrony krajobrazu*. Lublin: ZLPK.

Richling, A. & Solon, J. 1994. *Ekologia krajobrazu*. Warsaw: PAN.

Roberts, J. 1973. *Civilization. The Emergence of Man in Society*. Del Mar: CRM Books.

Robinson, E. 2007. Exxon Exposed. *Catalyst* Spring: 2–4.

Rogala, P. 2003. Podstawy nauki o jakosci. In T. Borys & A. Ptaszynska (eds), *Zarzadzanie jakoscia i srodowiskiem*: 83–84. Jelenia Gora: KZJiS.

Rogowski, R. 1992. Teoekologia. In C. Napiorkowski & W. Koc (eds), *Chronic, by przetrwac*: 31–45. Niepokalanow: Wydawnictwo Ojcow Franciszkanow.

Rok, B. 2004. *Odpowiedzialny biznes w nieodpowiedzialnym swiecie*. Warsaw: ARF, FOB.

Ropke, I. 1999. The Dynamics of Willigness to Consume. *Ecological Economics* 28: 399–420.

Ropke, I. 2001. New Technology in Everyday Life – Social Processes and Environmental Impact. *Ecological Economics* 38: 403–422.

Rosenkranz, G., Froggatt, A., Kreusch, J., Neumann, W., Appel, D., Diehl, P., Nassauer, O., Thomas, S. & Matthes, F. Chr. 2006. *Energia jadrowa, mit i rzeczywistosc. O zagrozeniach zwiazanych z energiq jadrowa i jej perspektywach w przyszlosci*. Warsaw: FHB.

218

Rugumayo, E. 1987. Key Issues in Environmental Education. In A.V. Baez, G.W. Knamiller & J.C. Smyth (eds), *The Environment and Science and Technology Educaton, Science and Technology Education and Future Human Needs, vol. 8*: 29–35. Oxford, New York, Bejing, Frankfurt, Sao Paulo, Sydney, Tokyo, Toronto: Pergamon Press.

Runc, J. 1998. *Ochrona srodowiska a konflikty spoleczne w Polsce*. Poznan: WSNHiD.

Roziewicz, M. & Wendlandt, J. 2004. *Pomniki historii*. Warsaw: KOBiDZ.

Ryden, L., Migula, P. & Andresson, M. 2003. *Environmental Science*. Uppsala: Baltic University Press.

Ryden, L. 1997. *Foundations of Sustainable Development. Ethics, Law, Culture, and the Physical Limits*. Uppsala. Uppsala Publishing House.

Sachs, I. 1990. Trwaly rozwoj, od koncepcji normatywnej do dzialania. *Ekonomista* 2–3: 358.

Sachs, W. 1993. *Global Ecology: an Arena of Political Conflicts*. London: Zed Books.

Sachs, J. 1994. *Poland's Jump to the Market Economy*. Cambridge: MIT.

Saint Marc, P. 1971. *Socialisation de la nature*. Paris: Stock.

Salay, J. 1997. *Energy, From Fossil Fuels to Sustainable Energy Resources*. Uppsala: The Baltic University Programme.

Scherr, S.J. 2001. People and Environment: What is the Relationship Between Exploitation of Natural Resources and Population Growth in the South? In J.M. Harris, T.A. Wise, K.P. Gallagher & N.R. Goodwin (eds), *A Survey of Sustainable Development. Social and Economic Dimensions. Frontier Issues in Economic Thought*: 138. Washington, Covelo, London: Island Press.

Schumacher, E.F. 1973. *Small is Beautiful, a Study Or Econimics as if People Mattered*. London: Blond & Birggs.

Schweitzer, A. 1993. The Ethics of Reverence for Life. In S.J. Armstrong & R.G. Botzler (eds), *Environmental Ethics. Divergence and Convergence*: 342–346. New York: McGraw-Hill Inc.

Schnoor, J.L. 2003. An Environmental Challenge. *Environmental Science and Technology* 1 April: 119 A.

Seidel, J. 2006. *Sustainablilty in Civil Engineering*. Edinburgh: Heriot-Watt University.

Seidler, C. & Peschke, G. 2004. Ekologiczny, ekonomiczny i spoleczny wymiar proekologicznego zarzadzania przedsiebiorstwem. In M. Kramer, M. Urbaniec & A. Krynski (eds), *Miedzynarodowe zarzadzanie srodowiskiem*: 1. Warsaw, C.H. Beck.

Seuring, S. 2008. Sustainability and Supply Chain Management. *Journal of Cleaner Production* 2: 1–7.

Shephard, P. & McKiney, D. 1969. *The Subversive Science: Essays Toward an Ecology of Man*. Boston: Houghton Mifflin Co.

Shrader-Frechette, K. 2006. Individualism, Holism, and Environmental Ethics. In N. Haenn & R.R. Wilk (eds), *The Environment in Anthropology. A Reader in Ecology. Culture and Sustainable Living*: 336. New York, London: New York University Press.

Simmons, I.G. 2006. Normative Behavior. In N. Haenn & R.R. Wilk (eds), *The Environment in Anthropology. A Reader in Ecology. Culture and Sustainable Living*: 53. New York, London: New York University Press.

Skalski, K. 2007 *Rewitalizacja na blokowisku*, http://www.fr.org.pl/uploaded_docs/Rewitalizacja%20na%20blokowisku_na20%www.doc [30.06.2009].

Skolimowski, H. 1993. *Filozofia zyjaca, eko-filozofia jako drzewo zycia*. Warsaw: Pusty Oblok.

Skolimowski, H. 2007. Glowne tezy. *Zielone Brygady* 2(223).

Skowronski, A. Ekofilozoficzny wymiar koncepcji zrownowazonego rozwoju. In A. Pawłowski (ed.), *Filozoficzne, spoleczne i ekonomiczne zrownowazonego rozwoju*: 140–141. KIS, Lublin 2004.

Skowronski, A. 2008. A Civilization Based on Sustainable Development: It's Limits and Prospects. *Sustainable Development* 2(16): 117–125.

Sky News. 2011. *News between 11 and 17 March*.

Smyth, J.C. 1987. Key Environmental Issues and Formal Education. In A.V. Baez, G.W. Knamiller & J.C. Smyth (eds), *The Environment and Science and Technology Educaton, Science and Technology Education and Future Human Needs, vol. 8*: 47–50. Oxford, New York, Bejing, Frankfurt, Sao Paulo, Sydney, Tokio, Toronto: Pergamon Press.

Social Exclusion: Rhetoris, Reality, Responses. 1995. Geneva: International Labor Organization.

Soto, H. De. 2000. *The Mystery of Capital, Why Capitalism Triumphs in the West and Fails Everywhere Else*. New York: Basic Books.

Sörlin, S. 1997. *The Road Towards Sustainability. A Historical Perspective*. Uppsala: Uppsala Publishing House.

Sörlin, S. 2003. The Story of Easter Island. In L. Ryden, P. Migula & M. Andersson (eds), *Environmental Science*. Uppsala: The Baltic University Press.

Spangenberg, J.H. 2005. Economic Sustainability of the Economy: Concepts and Indicators. *International Journal of Sustainable Development* 1–2(8): 48–49.

Stenmark, M. 1997. Humans and the Value of Nature. In L. Ryden (ed.), *Foundations of Sustainable Development. Ethics, Law, Culture, and the Physical Limits*: 20. Uppsala: The Baltic University Press.

Steward, J. 2006. The Concept and Method of Cultural Ecology. In N. Haenn & R.R. Wilk (eds), *The Environment in Anthropology. A Reader in Ecology. Culture and Sustainable Living*: 5–9. New York, London: New York University Press.

Stepien, A.B. 1989. *Wstep do filozofii*. Lublin: KUL.

Stokes, B. 1981. *Helping Ourselves. Local Solutions to Global Problems*. New York, London: W.W. Norton & Company.

Strahl, J. 1997. *Sustainable Industrial Production. Waste Minimization, Cleaner Technology and Industrial Ecology*. Uppsala: Uppsala Publishing House.

Strahl, J., Sorgaard, M. & Lorentzen, B. 1997. Environmental Management Options in the Firm. In J. Strahl (ed.), *Sustainable Industrial Production. Waste Minimization, Cleaner Technology and Industrial Ecology*: 23–24. Uppsala: Uppsala Publishing House.

Striebig, B., Jantzen, T., Rowden, K., Dacquisto, J. & Reyes, R. 2006. Learning Sustainability by Design. *Environmental Engineering Science* 3(23): 439–440.

Stodulski, W. 2004. Ekologiczna Reforma Fiskalna jako element reformy systemu finansow publicznych, instrumenty rozwoju. *EkoProfit* 1(69): 38–48.

Stola, W. 1984. An Attempt at a Functional Classification or Rural Areas In Poland. A Methodological Approach. *Geographia Polonica* 50: 113–129.

Strandberg, H. 1992. Chinese Coal Darkens the Sun. *Enviro* 14: 14–16.

Strasfeld, R. 2001. Regionalna ekonomia jako przestrzenny przyczynek orientacyjny dla zrownowazonego rozwoju. *Problemy Ekologii* 2(5): 61.

Streeten, P. 1998. Globalization: Threat or Salvation? In Bhalla A.S., *Globalization, Growth, and Maginalization*: 13–46. New York: St Martin's Press.

Sully, J. & Hill, N. 2006. *EMEP/CRINAPIR Emission Inventory Guidebook 2006*. EEA. http://reports.eea.europa.eu/EMEPCORINAIR4/en/page004.html [30.06.2007].

Szaniawska, D. & Szaniawski, A.R. 1995. Zrownowazony rozwoj. *Ekonomia i Srodowisko* 2(7): 45–54.

Sztombka, W. 1991. *Etyka badan naukowych. Filozoficzna struktura problemow. Lodz*: UL.

Sztombka, W. 1994. Hansa Jonasa etyka odpowiedzialnosci. *Ethos* 25–26: 124–134.

Sztumski, W. 1999. *Moment ekologiczny i postawa ekologiczna*. Bydgoszcz: Akademia Bydgoska.

Sztumski, W. 2000. Przejawy kryzysu ekologicznego w srodowisku spolecznym. In A. Papuzinski (ed.), *Polityka, ekologia, kultura. Spoleczne przeslanki kryzysu ekonomicznego*: 122. Bydgoszcz: WSP.

Sztumski, W. 2006. Idea zrownowazonego rozwoju a mozliwosci jej urzeczywistnienia. *Problemy Ekorozwoju/Problems of Sustainable Development* 2(1): 73–76.

Sztumski, W. 2009. The Mythology of Sustainable Development. *Problemy Ekorozwoju/ Problems of Sustainable Develkopment* 2(4): 13–24.

Szulczewska, B. & Cieszewska, A. 2006. Uklad przyrodniczy obszaru metropolitalnego. In S. Kozlowski (ed.), *Zywiolowe rozprzestrzenianie sie miast*: 57. Bialystok, Lublin, Warsaw: EiS.

Szafer, W. & Michajlow, W. 1973. *Ochrona przyrodniczego srodowiska czlowieka.* Warsaw: PWN.

Szafranski, A.L. 1993. *Chrzescijanskie podstawy ekologii.* Lublin: EkoKul.

Szymanski, W. *Etyka ekologiczna*, unpublished article.

Slipko, T. 1985. Podstawy etyki srodowiska naturalnego. *Chrzescijanin w swiecie* 4: 55–56.

Slipko, T. 1992. Czlowiek elementem przyrody ze stanowiska inkluzjonizmu i ekskuzjonizmu. In L. Pawłowski & S. Zieba (eds), *Jakiej filozofii potrzebuje ekologia*: 10–17. Lublin: Politechnika Lubelska.

Slipko, T. & Zwolinski, A. 1999. *Rozdroza ekologii.* Cracow: WAM.

Swiatek, L. & Charytonowicz, J. 2006. Przemyslec zrownowazone budowanie. *Recykling* 10(70): 30.

Swierczek, Z. 1990. *Ekologia – kosciol i sw. Franciszek.* Cracow: WSD.

Tengstrom, E. & Thynell, M. 1997. *Towards Sustainable Mobility. Transporting People and Goods in the Baltic Region.* Uppsala: Ditt Tryckeri.

Thomas, G., Bell, P. & Baum, A. 2004. *Psychologia srodowiskowa.* Gdansk: WP.

Thomas, V., Theis, T., Lifset, R., Grasso, D., Kim, B., Koshland, C. & Pfahl, R. 2003. Industrial Ecology: Policy Potential and Research Needs. *Environmental Engineering Science* 1(20): 1–9.

Thornblad, H. 1992. Doubts on Early HCFC Phase-out. *Enviro* 13: 24–25.

Tibbs, H. 1993. *Industrial Ecology, An Environmental Agenda for Industry.* Emeryville: Global Business Network.

Tiberg, N. 1992. *The Prospect of a Sustainable Society.* Uppsala: Uppsala Publishing House.

Tinbergen, J. 1976. *Rio Report: Reshaping the International Order.* New York: Dutton.

Timoshenko, A. & Berman, M. 1996. The United Nations Environment Programme and the United Nations Development Programme. In J. Werksman (ed.), *Greening International Institutions*: 38–54. London: Earthscan.

Tjallingii, S.P. 1995. *Ecopolis – Strategies for Ecologically Sound Urban Development.* Leiden: Backhuys Publishers.

Tobera, P. 1988. *Kryzys srodowiska – kryzys spoleczenstwa.* Warsaw: LSW.

Toffler, E. 1980. *The Third Wave.* New York: Bantam.

Toffler, E. & Toffler, H. 1995. *Creating a New Civilization: The Politics of the Third Wave.* Paducah: Turner Pub., Nashville.

Tourism and Environment.1994. *Enviro* 17.

Toynbee, A. & Toynbee, A. 1948. *Civilization on Trial.* Oxford: Oxford University Press.

Trzaskowska, E. & Sobczak, K. 2006. Jak chronic krajobraz wsi podmiejskich przed rozlewaniem sie miast na przykladzie okolic Lublina. In S. Kozlowski (ed.), *Zywiolowe rozprzestrzenianie sie miast*: 323. Bialystok, Lublin, Warsaw: EiS.

Trzeciak, P. 1957. *Zagadnienia ochrony zabytkow architektury.* Warsaw: PTTK.

Turner, M.G., Gardner, R.H. & O'Neill, R.V. 2001. *Landscape Ecology in Theory and Practise. Pattern and Process.* New York: Springer.

Tweed, Ch. & Sutherland, M. 2007. Built Cultural Heritage and Sustainable Urban Development. *Landscape and Urban Planning* May.

Tyburski, W. 2002. Komponenty kultury ekologicznej. In J.M. Dolega (ed.), *Podstawy kultury ekologicznej*: 22. Warsaw: UKSW.

Tyburski, W. 2002. Praktyczny wymiar etyki srodowiskowej. In J.W. Czartoszewski (ed.), *Etyka srodowiskowa wyzwaniem XXI wieku*: 106–110. Warsaw: Verbinum.

221

Tyburski, W. 2003. Problematyka jakosci zycia w perspektywie etyki srodowiskowej. In J. Tomczyk-Tolkacz & A. Ptaszynska (eds), *Jakosc zycia w perspektywie nauk humanistycznych, ekonomicznych i ekologii*: 330–343. Jelenia Gora: KZJiS.

Tyburski, W. 2004. O niektorych aksjologicznych przeslankach zrownowazonego rozwoju. In A. Pawłowski, (ed.). *Filozoficzne, spoleczne i ekonomiczne uwarunkowania zrownowazonego rozwoju*: 213–222. Lublin: KIS.

Tyburski, W. 2008. Origin and Development of Ecological Philosophy and Environmental Ethics and Their Impact on the Idea of Sustainable Development. *Sustainable Development* 2(16): 100–108.

Udo, V. & Pawłowski, A. 2010. Human Progress Towards Equitaible Sustainable Development: A Philosophical Exploration. *Problemy Ekorozwoju/Problems of Sustainable Development* 2(5): 23–44.

Ukidwe, N.U. & Bakshi, B.R. 2005. Flow of Natural versus Economic Capital in Industry Supply Networks and Its Implications to Sustainability. *Environmental Science & Technology* 24(39): 9759–9769.

Universal Encyclopedia. 1984, 2000. PWN, Warsaw.

UNIZAR. 2011. *The European Union Constitution*. http://www.unizar.es/euroconstitucion/ Treaties/Treaty_Lisbon_Rat.htm [15.03.2011].

U.S.V. Smithfield Foods Inc. 2007. *Civil Action* 1:03-CV-00434.

U'Thant S. 1969. *The Problems of Human Environment*. New York: UN.

Valaskakis, K., Sindell, P.S., Smith, J.G. & Fitzpactrick-Martin, J. 1981. *The Conserver Society. A Workable Alternative for the Future*. New York: Harper and Row.

VanOverbeke, D. 2004. *Water Privatization Conflicts*. http://academic.evergreen.edu/g/grossmaz/ VANOVEDR/ [1.03.2009].

Vanclay, F. & Bronstein, D.A. 1995. *Environmental and Social Impact Assessment*. Chichester, New York, Brisbane, Toronto, Singapour: Wiley & Sons.

Verbruggen, A. 2000. How Scientific Research Can Contribute to Advance a Policy of Sustainable Development: an Examination of the "Communication Gap" Separating Policy and Research. In *Sustainable Development. Towards a Sustainable Dialogue Between Science and Policy*: 22–29. Brussels: Federal Office for Scientific, Technical and Cultural Affair.

Vermeulen, W.J.V. & Ras, P.J. 2006. The Challenge of Greening Global Product Chains: Meeting Both Ends. *Sustainable Development* 14: 245–256.

Vince, G. 2002. Safety Violations Shut Dutch Nuclear Reactor. *New Scientist Tech* 4 Feubruary.

Viscusi, W.K. 1992. *Pricing Environmental Risks*. St. Louis: Washington University.

Vosti, S.A. & Reardon, T. 1997. *Agroecological Perspective*. Baltimore: John Hopkins University Press.

Wackernagel, M. & Rees, W.E. 1996. *Our Ecological Footprint: Reducing Human Impact on the Earth*. Gabriola Island: New Society Publishers.

Walczak, M., Radziejowski, J., Smogorzewska, M., Sienkiewicz, J., Gacka-Grzesikiewicz, E. & Pisarski, Z. 2001. *Obszary chronione w Polsce*. IOS: Warsaw.

Wallerstein, I. 1999. *The End of the World As We Know It: Social Science for the Twenty-first Century*. Minneapolis: University of Minnesota Press.

Wasilewski, M. 2006. Strategie i narzedzia rozwoju zrownowazonego – zmiany wzorcow produkcji konsumpcji. *Czystsza Produkcja i Eko-zarzadzanie* 4(39): 17.

Water, Safe, Strong and Sustainale. Vision on European Water Supply and Sanitation in 2030. 2004. Brussels: Water Supply and Sanitation Platform.

WCED. 1987. *Our Common Future. The Report of the World Commission on Environment and Development*. New York: Oxford University Press.

Weigle, A. 1996. A Strategy for Biodiversity Protection in Poland. In A. Breymeyer & R. Noble (eds), *Biodiversity Conservation in Transboundary Protected Areas*: 70–79. Washington: NAP.

222

Weis, P. & Bentlage, J. 2006. *Environmental Management Systems and Certification*. Uppsala: The Baltic University Press.

Weizsäcker, E.U. von, Lovins, A.B. & Lovins, L.H. 1997. *Factor Four: Doubling Wealth, Halving Resource Use – The New Report to the Club of Rome*. London: Earthscan: London.

Weizsäcker, E.U., Hardgroves "Charlie", K., Smith, M.H., Desha, C. & Stasinopoulos, P. 2009. *Transforming Global Economy Through 80% Improvements in Resource Productivity*. London: Earthscan.

Werksman, J. 1996. *Greening International Institutions*. London: Earthscan.

White, L. 1996. The Historical Roots of Our Ecological Crisis. *Science* 155: 1204–1207.

Wernick, I.K., Herman, R., Govind, S. & Ausubel, J.H. 2001. Materialization and Dematerialization. Measures and Trends. In J.M. Harris, T.A. Wise, K.P. Gallagher & N.R. Goodwin (eds), *A Survey of Sustainable Development. Social and Economic Dimensions. Frontier Issues in Economic Thought*: 198–201. Washington, Covelo: Island Press.

The Wilderness Act Handbook. 2004. Washington: The Wilderness Society.

Wilk, R.R. 2006. The Ecology of Global Consumer Culture. In Haenn, N. & Wilk, R.R. (eds.), *The Environment in Anthropology. A Reader in Ecology. Culture and Sustainable Living*: 418–429. New York, London: New York University Press.

Williams, K. & Dair, C. 2007. A Framework of Sustainable Behaviours that Can be Enabled Through the Design of Neighbourhood-Scale Developments. *Sustainable Development* 15: 160–173.

Williams, K. & Dair, C. 2007. What is Stopping Sustainable Building in England? Barriers Experienced by Stakeholders in Delivering Sustainable Development. *Sustainable Development* 15: 142.

Wilson, E.O. 2001. Nature Matters. *American Journal of Preventive Medicine* 3(20): 241.

Wise, T.A. 2001. Global Perspecitves: The North/South Imbalance. In J.M. Harris & N.R. Goodwin (eds), *A Survey of Sustainable Development. Social and Economic Dimensions. Frontier Issues in Economic Thought*: 81. Washington, Covelo, London: Island Press.

Wise, T.A. 2005. Economics and Sustainability: The Social Dimension. *International Journal Sustainable Development* 1–2(8): 50–51.

Wodziczko, A. 1948. *Na strazy przyrody. Wiadomosci i wskazania z dziedziny ochrony przyrody*. Cracow: PROP.

Wolfram, K. 2005. Zielone Pluca Polski – jako program wdrozeniowy zrownowazonego rozwoju regionu polnocno-wschodniej Polski. In B. Poskrobko & S. Kozlowski (eds), *Studia nad zrownowazonym* rozwojem: 304–315. Bialystok, Warsaw: KCiS.

Wojciechowski, I. 1997. Ekologia jako nauka stosowana w ochronie przyrody i ochronie srodowiska. In T. Puszkar & L. Puszkar (eds), *Wspolczesne kierunki ekologii, ekologia behawioralna*: 21–29. Lublin: UMCS.

Wojciechowski, K.H. 2004. Niemierzalne skladniki krajobrazu. In M. Kucharczyk (ed.), *Wspolczesne problemy ochrony krajobrazu*: 17–23. Lublin: ZLPK.

Wojciechowski, K.H. 2006. Ubran Sprawl – naturalne, czy wymuszone stadium sukcesji krajobrazowej? In S. Kozlowski (ed.), *Studia nad zrownowazonym rozwojem*: 72–79. Bialystok, Lublin, Warsaw: EiS.

Wojtas, M. 2005. Rola i funkcje organizacji miedzynarodowych we wspolczesnej gospodarce swiatowej. In B. Mucha-Leszko (ed.), *Wspolczesna gospodarka swiatowa, glowne centra gospodarcze*: 345–346. Lublin: UMCS.

Wolanski, N. 1992. Jakiej podstawy naukowej potrzebuja dzialania proekologiczne? Ochrona, uzytkowanie i ksztaltowanie przestrzeni zycia czlowieka. In L. Pawłowski & S. Zieba (eds), *Jakiej filozofii potrzebuje ekologia*: 106. Lublin: Politechnika Lubelska.

Wolanski, N. & Siniarska, A. 2001. Wychowanie do srodowiska w swietle ekologii czlowieka. In J.W. Czartoszewski (ed.). *Edukacja ekologiczna na progu XXI wieku*: 63–86. Warsaw: Verbinum.

Wolanski, N. & Siniarska, A. 2002. Kultura jako niebiologiczny sposob przystosowania do srodowiska (wychowanie i ksztalcenie jako przejaw kultury). In J.M. Dolega (ed.), *Podstawy kultury ekologicznej*: 33–42. Warsaw: KCiS.

Woodward, L. & Lampkin, N. 1996. Organic Farming. In R. Jones, M. Summers & E. Mayo (eds), *Sustainable Agriculture, Economic Alternatives for Eastern Europe* 3: 22–34. London: New Economics Foundation.

World Bank. 1994. *Making Development Sustainable. The World Bank Group and the Environment*. Washington: World Bank.

World Charter for Nature. 1982. New York: UN. http://www.un.org/documents/ga/res/37/a37r007.htm [30.06.2007].

World Conservation Strategy. Living Resource Conservation for Sustainable Development. Prepared by the International Union for Conservation of Nature and Natural Resources. 1980. New York: UCN. http://www.iucn.org/dbtw-wpd/edocs/WCS-004.pdf [30.06.2007].

World Population to 2300. 2004. New York: UN.

World Resources Institute. 2005. *Ecosystem and Human Well-Being. A Report of the Millennium Ecosystem Assessment*. Washington: Island Press.

Wrzosek, S. 1993. Ochrona prawna srodowiska przyrodniczego i przestrzeni. In B. Prandecka (ed.), *Interdyscyplinarne podstawy ochrony srodowiska przyrodniczego*: 67–80. Wroclaw, Warsaw, Cracow: Ossolineum.

Zaborska, K. 2007. *Osiedle za murem – bezpieczne domy szczesliwych ludzi?* http://www.sztukakrajobrazu.pl/w6zaborska.htm [30.VI.2007].

Zbicinski, I. & Stavenuiter, J. 2006. *Product Design and Life Cycle Assessment*. Uppsala: The Baltic University Press.

Zbicinski, I., Stavenuiter, J., Kozlowska, B. & Coevering, H. van de. 2007. *Product Design and Life Cycle Assessment*. Uppsala: The Baltic University Press.

Zieba, S. 1992. Idea natury w XVII w., aspekt ekologiczny. *Colloquium Salutis* 23–24: 363–368.

Zieba, S. 1995. *Konferencje ekologiczne*. Lublin: KUL.

Zieba, S. 1996. Technika a bezpieczenstwo ekologiczne. *Czlowiek i Przyroda* 5: 68.

Zieba, S. 1996. W poszukiwaniu podstaw odpowiedzialnosci za przyrode. *Czlowiek i Przyroda* 4:85.

Zimbardo, P.G. & Ruch, F.L. 1996. *Psychologia i Zycie*. Warsaw: PWN.

Zukowska, H. 1996. *Ekonomiczne aspekty ochrony srodowiska naturalnego*. Lublin: UMCS.

Zwawa, A. 2005. Lokalna zywnosc – czym jest i czym nie jest, alterglobalizacja na talerzu. *Ekoprofit* nr 3(75): 89–91.

Zychowicz, D. 1998. Globalne zagrozenia ekologiczne a zasada sprawiedliwosci miedzygeneracyjnej. In B. Poskrobko (ed.), *Sterowanie ekorozwojem*: 152. Bialystok: PB.

Zylicz, T. 1991. Environmental Reform in Poland, Theories Meet Reality. In Z. Bochniarz & R. Bolan (eds), *Designing Institutions for Sustainable Development: a New Challenge for Poland*: 267–273. Minneapolis, Bialystok: BTU.

Zylicz, T. 1993. Ekonomiczne mechanizmy i instrumenty ochrony srodowiska. In B. Prandecka (ed.), *Interdyscyplinarne podstawy ochrony srodowiska przyrodniczego*: 213–229. Wroclaw, Warsaw, Cracow: Ossolineum.

Zylicz, T. 1997. *Ecological Economics, Markets, Prices and Budgets in a Sustainable Society*. Uppsala: Baltic University Programme.

Yardley, J. 2007. Under China's Booming North, the Future is Drying Up. *New York Times* 28 September.

Young, J.Z. 1971. *The Study of Man*. Oxford: Oxford University Press, Oxford.

ADDITIONAL INFORMATION FROM THE INTERNET SERVICES

AFC. http://www.afdb.org/ [30.12.2010].
AIA2030 Challenge. http://www.aia.org [31.12.2009].
Allinance of Relligion and Conservation. http://www.arcworld.org/about.asp?pageID=2#86 [30.05.2009].
CDIAC. http://cdiac.ornl.gov/trends/emis/em_cont.htm [30.06.2010].
The Club of Rome. http://www.clubofrome.org [30.10.2010].
Earth Charter Initiative. http://www.earthcharter.org/ [30.06.2008].
Earth Summit 2002. http://www.earthsummit2002.org [30.06.2008].
EBRD. http://www.ebrd.com/ [30.12.2008.].
ECOSOC. http://www.un.org/docs/ecosoc/ [30.12.2008].
EEAC Network. 2004. http://www.eeac-net.org/ [30.06.2007].
Europa – The European Union On-line. http://europa.eu/scadplus/treaties/singleact_en.htm [30.02.2011].
European Environmental Agency. http://www.eea.europe.eu [30.02.2011].
EIA. http://www.eia.doe.gov/emeu/international/reserves.html [30.12.2008].
EMAS. http://ec.europa.eu/environment/emas/ news/index_en.htm#224 [30.01.2008].
DESA. http://www.un.org/desa [30.06.2009].
FAO. http://www.fao.org/ [30.06.2009].
Free Trade and Organization. http://www.globalissues.org/TradeRelated/Seattle.asp [3.1.2011]
GEF. Global Environmental Facility. http://www.ebrd.com/ [30.12.2009].
Green Party of Aotearoa. http://www.greens.org.nz/about/history.htm [30.06.2008].
Habitat. http://www.unhabitat.org [30.06.2008].
HIPCI. http://www.imf.org/external/np/exr/facts/hipc.htm [31.12.2010].
Hr-Net, Hellenic Resources Network, http://www.hri.org/docs/Rome57/ [30.06.2009].
Human Rights Watch. http://hrw.org/ [10.03.2011].
ICLEI. http://www.ilcei.org/ [30.06.2009].
IDA. http://web.worldbank.org/ida/ [31.12.2009].
IHP's Primary Objectives. http://www.unesco.org/water/ihp/ [30.06.2008].
International Institute for Sustainable Development. http://www.iisd.org/rio+5/ [30.06.2009].
IPCC. Intergovernmental Panel on Climate Change. http://www.ipcc.ch/ [31.12.2009].
ISO. http://www.iso.org [10.03.2011].
IUCN. http://www.iucn.org/ [10.03.2011].
Kinsale Challenge. EEAC, http://www.eeac-net.org/download/EEAC-KinsaleChallenge_17-4-04.pdf [30.06.2008].
LEED. http://www.usgbc.org/ [31.12.2010].
The Library of Congress. http://memory.loc.gov/ammem/gmdhtml/yehtml/yeabout.html [30.06. 2009].
The Life and Legacy of Rachel Carson. http://www.rachelcarson.org [stan z 30 VI 2007 r.].
MaB. http://www.unesco.org/mab/ [30.01.2011].
Millenium Declaration. http://www.un.org/millennium/declaration/area552e.htm [30.09.2009].
Millennium Development Goals Report. http://www.un.org/millenniumgoals/documents.html [30.09.2009].
Millenium Summit. http://www.un.org/millennium/ declaration/area552e.htm [30.06.2010].
NATURA 2000. http://natura2000.eea.europa.eu/ [10.12.2011].
The Natural Step. Http:///www.naturalstep.org [10.03.2011].
PICTURE. Pro-active Management of the Impact of Cultural Tourism Upon Urban Resources and Economies. http://wwww.picture-project.com [31.12.2010].

Public Citizen, Protecting Health, Safety & Democracy. http://www.citizen.org/cmep/foodsafety/eu/eucafo/pol/index.cfm [30.06.2009].
Regional Greenhouse Gas Initiative, an Initiative of the Northeast & Mid-Atlantic States of the U.S. http://www.rggi.org/ [12.03.2011].
SCOPE. http://www.icsu-scope.org/ [30.06.2009].
Sharing Sustainable Solutions. http://www.sharingsustainablesolutions.org/ [14.03.2011].
Sierra Club. http://www.sierraclub.org [31.12.2010].
Silent Spring Institute. http://www.silentspring.org [30.06.2009].
UN. http://www.un.org/ [30.09.2010].
UNDP. http://www.undp.org/ [30.09.2010].
UNEP. http://www.unep.org/ [30.09.2010].
UNESCO. http://www.unesco.org [10.01.2011].
UNICEF. http://www.unicef.org [10.03.2011].
UNIDO. http://www.unido.org/ [30.06.2009].
The U.S. Conference of Mayors, Mayors Climate Protection Center. http://usmayors.org/ climateprotection/listofcities.htm [10.03.2011],
WHO. http://www.who.int/about/en [stan z 30 VI 2007].
WH UNESCO. http://whc.unesco.org/en/list/ [30.12.2010].
Wikipedia. http://ww.wikipedia.org [15.03.2011].
WMO. http://www.wmo.ch [stan z 30 VI 2007 r.].
World Crude Oil and Natural Gas Reserves. http://www.eia.doe.gov/emeu/international/reserves.html [10.03.2011].
World Watch Institute. http://www.worldwatch.org/ [30.12.2010].

Index

Acid Rains 146
Agenda 21 21–22
Agricultural Environment 95–98, 111–112
Alter-globalism 186
Ancient Rome Civilization 102
Anti-globalism 186
Anthropocentrism
  Modertate Anthropocentrism 60
  Trap of Anthropocentrism 61

Backcasting 71
Biocentrism 61
Biogas 159
Biomass 156–157
Bovine Spongiform Encephalopathy
  (BSE) 97

Capital
  Financial capital 126
  Human capital 126
  Natural capital 126
  Physical capital 126
  Social capital 126
Categories of protected areas (IUCN) 87
Category of Development 37–38
Civilizational Fear 71
Cleaner Production 149
Club of Rome 13
Club of Rome Reports 13–15, 24–25
Conserver Society 47
Consumer Society 64
Convention on Biological Diversity
  22–23
CORINE 88–89
Culture 104–105
CULTURE 2000 107
Cultural Ecology 82
Cultural Landscape 105–107

Decade of Education for Sustainable
  Development 77
Declaration "Ethique et spiritualité de
  l'environment" 72
Deep Ecology 62–63
Democracy (Principles of) 171

Demographic Crisis 118–120
Discounting 131

Earth Charter 2000 26, 51
Earth Summit in Johannesburg 27–28
Earth Summit in Rio de Janeiro 21–24
Ecocentrism 61
Eco-development 13, 19
Eco-efficiency 149
Eco-labels 137
Ecological Attitude 77–78
Ecological Crisis 58
Ecological Decalogue 73
Ecological Economy 125–128
Ecological Education 77
Ecological Education (Levels of) 77–81
Ecological Farming 95–96
Ecological Footprint 179–180
Ecological 'Global Marshall Plan'
  167–168
Ecological Sin 74
Ecologic Era (Principles of) 53
ECONET 89
Proposals of Economic Instruments 139
Economic Instruments (OECD) 131–132
Economic Principles in Environmental
  Protection 127
Economics 123
Eco-philosophy (Issues of) 57–58
Eco-Philosophy (Standpoints) 60–62
Ecosophy 62–63
Ecosystem 86
  Typology of Ecosystems 86
EMAS 135–137
Energy
  Biogas 159
  Biomass 156–157
  Geothermal Energy 159
  Hydropower 157
  Nuclear Power 153–156
  Renewable Energy Sources 156
  Solar Energy 158
  Thermal Power 151–152
  Total Electricity Capacity 152
  Wind Power 158

Environmental Accounts 129
Environmental Audits 137
Environment Economic Evaluation 128–129
Environmental Engineering 150
Equator Principles 142
Environment Protection 85–86
Environment Protection (Motives for) 2
Individual Contribution to Environment Protection 81
Environmental Reports 135
Ethics
  Christian Ethics 72–76
  Environmental Ethics 59
  Ethics of Responsibility 69–71
EU Climate Package 33
EU Environmental Action Programmes 30–32
EU Scenarios of the Political Future 170
EU Strategy of Sustainable Development 34–35
Euthyphronics 101
Eutrophication 146

Factor Five 25–26
Factor Four 24–25
Factory Farming 96–98
Forest Principles 23
Framework Convention on Climate Change 23–24

Generation 68–69
Globalization
  Cultural Globalization 182
  Economic Globalization 182
  Egoistic Globalization 182–183, 186–187
  Inclusive Globalization 182–183, 186–187
Green Chemistry 150
Green Lungs (Poland) 122–123
Greenhouse Effect 146–147

Healthy Cities (WHO) 113–114
Hierarchy of a Biosphere 86
Historical Monuments (List of, Poland) 108–110
Hunter-gatherer Period 189
Human Development Index 120
Human Ecology 81–82
Human Population 118–120
Humanistic Geography 82
Human Poverty Index 120

Impacts of Environmental Problems (Short- and Long-term) 176
Imperative of Responsibility 69–70
Industrial Capitalism 124
Industrial Ecology 148
ISO Standards 135–136
Integrated Environmental EU Directives 34
Intergenerational Relations 69

Kyoto Protocol 137

Landscape
  Divisions of Landscapes 93–94
  Cultural Lanscape 104–111
Landscape Ecology 99
Law
  Administrative Law 160
  Civil Law 160–161
  Criminal Law 160–161
  Criteria of Legal Norms 161–162
  Enforceability of the Law 161
  Functions of the Law 161
  Natural Law 160
LCA 133–134
Lisbon Strategy 34–35
Living Planet Index 86

Maastricht Treaty 33
Malthus Doctrine 119
Management or Raw Materials and Production Processes (Strategies) 148
Millennium Declaration 26, 65
Millennium Development Goals Report 26
Millennium Summit 26

NATURA 2000 89–91
Nature Conservation 85–86
Needs 64
Needs (Division of, Maslow) 64

Opacities of Human Activity 60
Our Common Future 17–19
Our Future, Our Choice (EU Programme) 31–32
Ozone Hole 146, 172–173

Paradigms of Sustainable Development (Netherlands) 52
Personological Economy 143
Philosophical Audit 81–83
Polish Ecological Club (PKE) 16, 43
Polynesians 102–103

Population 20, 86, 118–120
Progress 37–38

Quality of Life 80

Regional Economics 187
Regress 38
Responsibility
   Contracted Responsibility 67
   Imperative of Responsibility 69–70
   Natural Responsibility 67
Responsible Business (Principles of) 134
Revolution
   Agricultural Revolution 189
   Industrial Revolution 191–192
   Information Technology Revolution
     192–193
   Modernisation Revolution 193
   Revolution of Efficiency 24–26
   Scientific Revolution 191
   Sustainable Development Revolution
     193–194
Rio Declaration 21, 24, 48

Seven Point Pledge 174
Shift Tax 132
Shock Doctrine 184–185
Single European Act 29, 33
Smog ('Los Angeles') 146
Smog ('London') 146
Social and Political Darwinism 185
Social Exclusion 114–115
Social Space 115–116
Spatial Order 94
Stockholm Conference 11–12
Stockholm Declaration 11–12
Strategy 166–168
Suburban Functional Categories 111–112
Sumerian Civilization 102
Sustainability
   Strong sustainability 127
   Weak sustainability 127

Sustainable Development Definitions
   39–45
Sustainable Development Indicators
   53–55
Sustainable Development Financial
   Security 138–139
Sustainable Development Objectives
   63–64
Sustainable Development Philosophical
   Assumptions 63
Sustainable Development Planes 45–47
Sustainable Development Principles
   47–53, 163
Sustainable Development Strategies
   (Formulation of) 166–167

Technology 145
Theo-ecology 74
Towards Sustainability (EU Progamme) 31
Turbo-capitalism 182

Urban Cracks 115
Urban Environment 98–100, 111–116
Urban Management 113
Urban Sprawl 99
U'Thant Report 10

Values
   Ecological Values 66
   Universal Values 65

World Charter for Nature 17, 26, 51
World Conservation Strategy 17
World Heritage Sites (List of, UNESCO)
   107–108
Worldwatch Institute 15, 28–29